cheating is a gift
man gives himself
— M. Burns

shirt

velcro lines
"flap"
with words/sets
underneath

M. Kennette   2018

BRAIN WAVES THROUGH TIME

12 Principles for
Understanding the Evolution of the
Human Brain and Man's Behavior

BRAIN WAVES
THROUGH TIME

12 Principles for
Understanding the Evolution of the
Human Brain and Man's Behavior

DR. ROBERT T. DeMOSS

PLENUM TRADE • NEW YORK AND LONDON

ISBN 0-306-46010-6

© 1999 Robert T. DeMoss
Plenum Trade is a Division of Plenum Publishing Corporation
233 Spring Street, New York, N.Y. 10013

10 9 8 7 6 5 4 3 2 1

A C.I.P. record for this book is available from the Library of Congress

CONTENTS

ACKNOWLEDGMENTS

To the extent this work has value, the credit must be given to many individuals.

Without the unwavering support and generosity of my mother, Edna Adams, this project would have gone no further than an idea.

Without the persistence of Rhonda J. Winchell of the Author's Agency, *Brain Waves* would have remained only a sketch.

The difficult task of taking a *very* rough draft to completion could not have occurred without the critical and helpful comments of Linda Greenspan Regan of Plenum. She has an amazing ability to help transform a concept into something more. Marvin Newman contributed to the success of this project in ways that he may never fully know. He is the most unique individual I have ever met and one of the smartest. He taught me that we "experts" ultimately know only a little about others. Marvin has served as a guide into the mysteries of other cultures and worlds that were beyond my wildest imaginings.

This book was made possible, in part, through the help of Michael J. Daugherty, whose indomitable spirit lives on after the demise of his body.

Aaron Brower contributed to this book in the very beginning, long before I had any thoughts of writing a book. His laughter lives on, making the world a slightly better place.

Without question, this project was a group endeavor, but any errors or omissions are entirely my own.

CHAPTER 1   THE PARADOX

It was 1978. A young man stood like a bronze statue in the middle of an isolated campground high in the Colorado Rockies. Earlier in the day, a concerned woman had called the sheriff's office requesting help for her mentally ill brother. The sheriff, in turn, called the local mental health center where I was a new employee and a recent graduate, having just received a master's degree in mental health counseling a few months earlier.

At the campground, I cautiously approached the disheveled man and explained why I was there. At first, the man just stood there. He showed no signs of curiosity about my presence, nor signs of embarrassment, even though he was soiled, his pants were soaked, and he smelled. Soon, he began talking about "the voices." He believed he heard the whisperings of aliens who were controlling his mind and telling him what to do.

Throughout our brief conversation, the man's facial expression and body posture remained immobile, with one exception. When I finished the evaluation, I explained to the man that he would be taken to the state mental hospital for further evaluation and treatment. At that moment, he seemed angry, but only briefly. Soon, his blank stare returned and his eyes lost their momentary spark. Shortly afterward, he entered the sheriff's car with no resistance, and I never saw him again.

Over two decades later, I can still picture the disturbed man as he stood there, staring off into a place most of us have never seen. I wonder what happened to him, although immediately after our visit, he was taken by the sheriff's deputy to the state hospital where he was diagnosed with *schizophrenia*, one of the most debilitating of all mental illnesses, one that is linked to abnormalities within the brain.[1] Shortly after being diagnosed, the man received medication and was released from the hospital.

On that day over two decades ago, I was certain of only one thing: My hard-won master's degree offered too little insight into the real-life dramas of the mentally disturbed.

Over twenty years later, another young man captured not only my attention, but that of the entire nation. On April 19, 1995, reportedly driven by "anger" and "alienation," Timothy McVeigh bombed a federal building in Oklahoma City, killing 168 people, wounding an additional 850.[2] Unlike the man in the campground, though, Timothy McVeigh was never diagnosed with schizophrenia nor any other debilitating mental illness—a fact that makes his behavior all the more baffling. After getting past the shock, grief, and outrage, an old question emerges: Why did he do it?

Our fascination with human behavior and the mind is not new. Throughout history, people have been puzzled, mystified, and even riveted by the actions of others. Hundreds of years before the time of Christ, the ancient Greeks were interested in unraveling the secrets of the mind. In about 400 B.C., the Greek physician Hippocrates was trying to explain depression by attributing it to an imbalance of body "humors," specifically, too much black bile.[3]

Literature, science, religion, philosophy, education, and politics all demonstrate our desire to know more about ourselves. But why, we might ask, do we have such an insatiable interest in human behavior? This entire book is devoted to this and other issues relating to behavior, and it explores this question in detail, with the hope of drawing some reasonable conclusions. In the meantime, though, there are other important issues that can be addressed here.

It was not until the 1970s that advanced techniques for studying live human brains emerged.[4] Nevertheless, a dearth of knowledge about the brain has *never* diminished our curiosity about human nature. In the absence of useful brain theories, people have always devised other explanations that satisfied their need to understand.

For most of human history, spiritual and religious explanations dominated theories of behavior.[5] In light of those theories, the brain often was forgotten or viewed as irrelevant, although its importance was not *completely* ignored. From time to time, a few scientists did recognize the brain's role in behavior.

In about A.D. 150, there was a Greek physician by the name of Galen who lived in Rome. He was a surgeon for the gladiators, and on the basis of his experience, he saw the types of behavior changes that resulted from brain injury. In spite of his observations, though, he could not dislodge a more popular theory of the time: Behavior was controlled by the heart, not the head.[6]

Historically, perspectives about behavior have largely been under the purview of religion and philosophy, but there are compelling arguments why behavior *should* be studied scientifically, and not solely by professionals. First, though, we must look at what is meant by "scientific."

The term "science" is often used to refer to a collection of facts, but more broadly, science involves a *process*, the steps required for studying the world (and universe) around us.[7] The first step begins with an observation, and for our purposes, we are concerned with the brain and behavior—both of which can be observed or studied scientifically.

Since you are reading this book, there is a good chance that you are curious about people. Curiosity, in turn, can be satisfied through scientific study, and one of the major tasks of scientists is to construct useful theories. Theories, in turn, are comprehensive statements that attempt to explain a phenomenon, like behavior. Psychology, which can be defined as the scientific study of behavior, has generated many theories pertaining to behavior.

Theories serve useful purposes. They tie together many observations and provide explanations about those observations. Historically, as noted, human behavior was attributed to spiritual forces, and the idea that behavior is caused by spirits is a theory but certainly not one that can be easily studied through scientific methods.

We often speak about theories as if they have been proved or disproved. However, a comprehensive theory, like the theory of evolution, is seldom completely verified or totally discredited. Rather, theories are general formulations that help us to seek answers by pointing us in certain directions.

Let us assume that we wanted to study the eating habits of a particular bug. Let us also assume that no one has ever observed this bug during the day. Thus, we have a partial theory already. We believe the insect is nocturnal. Our current interest pertains solely to the insect's eating habits. However, since we already hold a belief about the bug, our research into its eating habits will be influenced by our existing theory. In this instance, we are not trying to prove that the bug *is* nocturnal, because we can never prove that point conclusively. (There are too many bugs of this variety throughout the world to know that all of them are nocturnal. Thus, the belief that the bug is nocturnal will forever remain a theory.) Nevertheless, we will rely on our theory to study other aspects of the bug's behavior, like eating; and since we believe it is nocturnal, we will only search for it after dark.

As for the issue of spiritual forces and behavior, if we believe that evil spirits are the cause of abnormal behavior (as some people do), then we would have little interest in studying the brain in our attempt to understand abnormal behavior. Theories provide us with a starting assumption about some topic, and we rely on those assumptions to formulate specific questions that can be studied through scientific experimentation or some other means.

In contrast to theories, hypotheses are specific statements that are theoretical in nature, but ones that are limited in scope and can be tested. If I observed my neighbor arriving home late every night, and if I think I know why, I have a working hypothesis. I can test my hypothesis by making a prediction about the neighbor's future behavior and determining afterward if my prediction came true. My belief is that the neighbor comes home late every night to avoid a fight with his oldest son. Therefore, my hypothesis is that the neighbor will not come home late during those times when his son is away at college. Finally, if my prediction is borne out, I can further test the accuracy of my hypothesis by speaking with the neighbor.

In science, hypotheses (specific predictions) are tested through experiments, field studies, or scientific polls. A public elementary school might wish to know if one reading program works better for boys than for girls. To answer this question, the school's reading teachers might decide to conduct an experiment. At the beginning of the school year, they can randomly assign children to two different reading programs. (By randomly assigning the children, the teachers are able to counterbalance the effects of intelligence, previous reading ability, and so on.) Each class would then be given separate reading books and at the end of the year, the two classes would be tested to see if there was a difference between boys and girls as a function of different reading materials. If there was a difference, then girls and boys could be provided with different textbooks during the first years of school.

In the case of the reading experiment, the theory might be: "Boys and girls respond dissimilarly to the same reading program during the first three years of public school education." Because this statement is about boys and girls in general, it is theoretical—it is too broad to be proved or disproved easily, but its credibility can be inferred by a specific study with a testable hypothesis: "This curriculum is better for boys and this one is better for girls during the second year of public school education."

Testing specific hypotheses is a way of making inferences about the usefulness of a comprehensive theory. A complex theory might generate hundreds of experiments and hypotheses in order to test the theory's usefulness and its ability to explain and predict natural phenomena.

We are all scientists of sorts. We all hold theories about behavior and generate specific hypotheses all the time, and we even test our own hypotheses. If you are a parent, you have probably told your son or daughter, "If you stay up too late, you'll be too tired for school tomorrow." This statement is a hypothesis, because it makes a specific statement that can be tested scientifically. The hypothesis is tested the next day by determining if your son or daughter is still sleeping after the alarm has rung.

You also have a general theory that can never be fully proved or disproved, which is, "Children (in general) who stay up too late are too tired to get out of bed in time for school."

A news story about two California therapists provides a great illustration of how theories influence behavior. In May 1994, a jury in Napa, California, awarded a half million dollars to a plaintiff who convinced a jury that his daughter's memories of childhood abuse and incest were "planted" by two therapists. In the words of the plaintiff's attorney, "If they use nonsensical theories about so-called repressed memories to destroy peoples' lives, they will be held accountable." After the award was announced, a tearful therapist was quoted as saying, "Repressed memories are a reality."[8] This story illustrates an important point. *People behave on the basis of the assumptions and beliefs they hold about others,* whether those beliefs are theories or hypotheses.

In the foregoing instance, the therapists acted in accordance with their beliefs about behavior and mental systems. Unfortunately, their strong beliefs (a theory) did not protect them in a court of law. Far from being a "reality," as one therapist was quoted, "repressed memories" constitute a theory that is not fully consistent with new research. (In fairness to the therapists, though, their views are widely shared by thousands of professional psychotherapists, so it would not surprise me if the verdict was overturned on appeal.)

This story illustrates a second point: People's lives are affected when others act on the basis of theoretical beliefs. The problem of "faulty" theorizing can obviously have serious consequences for professionals, but psychotherapists are not the only ones who hold beliefs about why people behave as they do. As mentioned, *we all do*. Some interesting research on anger will further illustrate what I mean.

When we get angry at others, there is a good chance that we blame them for some offense.[9] Usually, the offense is considered intentional or, at the very least, careless. Notice, the key to anger is the *assumption* that an offense was intentional or carelessly inflicted. If we believe that an offense was accidental, we are far less likely to be angry, perhaps just annoyed, if that. Further, because anger and annoyance are two different emotions, we are inclined to act differently depending on which one we are feeling.

Imagine that you are standing in line for the theater and someone pokes you on the buttocks. You quickly turn around, only to see a small four-year-old boy who is wearing a little backpack, and it appears that the backpack was the cause of the poke since the child's back is turned toward you. Annoyance is far more likely than anger, and even annoyance may be too strong in this instance. On the other hand, assume that you receive the same poke, turn around, and there is a middle-aged man with a leer on his face. Anger is far more likely!

We act differently when we are annoyed from how we act (or wish to act) when we are angry. According to the research, when annoyed, people usually try to ignore the offense.[10] In turn, anger is more likely to result in a confrontation and at least a desire or thought of retaliation. (In truth, many people do not react in a physical manner, but many of us consider it.)

In the case of the small child or middle-aged man, anger or annoyance is based on an *assumption* about motive. In neither case do we know for certain why the individual behaved as he did. Our beliefs about the causes of behavior are hypotheses, and if we act on those beliefs, we are acting on the basis of an hypothesis or theory.

As individuals, we are not the only ones who hold hypotheses and theories about behavior and act on our beliefs. Groups of people, too, act in accord with theoretical views. Group beliefs are often molded into the laws of a society. As a general rule, like single individuals, society is more forgiving of accidents than premeditated acts, *even if an accident causes more harm than a premeditated act*. An accidental death as a result of a car accident may result in far less punishment than a bungled robbery attempt in which no injuries occurred.

As a society, we often punish offenders for their misdeeds, and we do so for many reasons: retribution, to make society a safer place for all of us, and to help offenders. Punishment is presumed to accomplish its goals in

avoid punishment ⎫
seek reward ⎭ behavior

two ways. First, it separates us from the offender (when punishment involves jail), and second, punishment is considered, by some, to be an effective way of changing an offender's behavior.

Now, suppose we look at punishment as a means of influencing behavior relative to the problem of drinking and driving. There are many years of research on punishment, so we can benefit from that knowledge.[11] In some circumstances, punishment can be very effective, but under other conditions, it does a poor job, if any, of changing behavior. With respect to drunk drivers, traditional ways of delivering punishment have been relatively ineffective. That is, about half of the people who have been caught for drinking and driving have been caught and punished before.[12]

drink
&
drive

How punishment affects a variety of behaviors is a useful topic to study. The answers we derive from this inquiry could have profound effects and meaning for society. This underscores the importance of a scientific study of human behavior. Such studies generate hypotheses and even theories that can have practical applications and perhaps lead to effective solutions to social problems. In short, better theories can lead to more effective solutions. This book, therefore, is pragmatic, if nothing else. It will help us generate theories about behavior that are in accord with scientific research.

Psychological explanations and theories of behavior have been criticized, partly because they have focused on *behavior* without addressing underlying brain or biological issues and partly because they have not always proved to be very useful. Nevertheless, we must depend on theories because of our need to address a variety of problems for which a complete understanding does not exist at present.

At present, we have to look beyond biological and brain research with animals and humans to consider the types of phenomena that interest so many of us, like how to help children read better, or how to help a neighbor get through a major disaster. We want to know about those behaviors that affect our daily lives, and we also want to understand the sensational stories of the day, like why Timothy McVeigh bombed the federal building in Oklahoma City. One reason why some psychological theories have been unsatisfactory, therefore, is that real-life problems are extremely complex, and it is hard for any theory to explain all behavior.

Even contemporary brain research, although impressive, still leaves many of our most pressing questions unanswered. Brain-based theories, for example, shed little light on why Timothy McVeigh felt justified in killing 168 people, even though he was driven by "anger" and "alienation." Because of the relative dearth of brain research that addresses problems like these, many disciplines continue to develop theories of behavior without the advantage of neurological information. Psychology, education, law enforcement, sociology, cultural anthropology, and even some biological specialties all have made, and continue to make, impressive discoveries about behavior without relying on brain research. Historically, and continuing to the present, the desire and need to understand ourselves, and respond to a variety of human problems, has long predated brain research, so there is clearly room for many approaches to studying the actions of people.

Given the complexity of human behavior, and the pressing need to know more about ourselves, I believe there is a solid middle ground from which we can build useful theories. We can take the best of two approaches and combine them. When careful observations of behavior are compared with newly emerging brain research, models of behavior can be generated that may be more powerful than theories that rely on behavior or brain research alone (though even then, not all behavior can be explained at present).

Even though powerful theories of behavior have been generated in the absence of brain research, once the technology became available to study the brain more effectively, it opened up a new era. It was the gold rush all over again, but this time, the new frontier was within the confines of the human skull. Very likely, during the past twenty-five years, more information has been uncovered about the brain than during all previous years of human existence. Ironically, though, the rising flood of brain research has failed to answer many of our most perplexing questions. Moreover, the amount of new information has itself produced some unexpected problems.

We should consider why so much information has not provided, as yet, more answers relative to the sheer volume being produced. There are many reasons for this paradoxical state of affairs.

First, brain imaging techniques are quite new and relatively rudimentary. Even the most modern imaging techniques can only study basic mental phenomena, like changes in metabolic processes when we think about some idea or repeat a word over and over. Many of the behaviors we take for granted, and their underlying mental processes, are unavailable for study by even the most sophisticated research equipment.

Second, most brain studies still rely on laboratory animals because animal and human brains are extremely similar in a number of important respects. However, when research is conducted on animals, it is not always clear to what extent the results can be applied to human behavior. In any event, research on animals continues to be the most common source of brain research, so we must rely on that data until better methods are developed for studying normal brain processes in healthy humans.

Third, although brain research is crucial for understanding brain biology and chemistry, we function within complex social environments. Studying the brain out of context can lead to erroneous conclusions, which is a problem that occurs often in theories of behavior.

We can now explore a more pivotal question. I mentioned that the flood of new brain research has created some unanticipated problems, and in my view, one problem is not going to be solved easily. *The sheer volume of information about the brain is one of the biggest impediments to understanding and applying that information.* The unabated explosion of facts about the brain has grown too rapidly for our capacity to assimilate it. Although computers can catalogue data, computers cannot sort through mounds of information and make useful inferences nor devise new theories. So, what can be done? A chance occurrence helps to answer this question.

In 1992, I was working as a counseling psychologist at a midwestern university. While there, one of my responsibilities was to supervise student interns pursuing their doctoral degrees in psychology, all of whom had obtained master's degrees, some of whom had a good deal of previous experience.

During a conversation with an intern who had specialized in working with adults who had been sexually or physically abused as children, I asked her: "Based on your experience, how might complex emotional memories be stored in the brain?" Acknowledged to be a professional in her field, she said, "How should I know?" Her answer was not the type we have come to expect from experts, but I hasten to add, I was her supervisor and had no more idea than she did.

At that time, and continuing to the present, most counseling and psychotherapy models do not include brain research. In fact, when I first embarked on this project, I purchased a standard textbook that is used in graduate school to teach various psychological theories, and the word "brain" is not even indexed.[13] How can that be?

First and foremost, most psychotherapy models were developed *before* relevant brain research became available. By "relevant" I mean research that could shed light on the types of problems that are normally the focus of psychotherapy, like strained marital relationships, troubles with the boss, problems relating to childhood experiences, chronic loneliness, thoughts of suicide, and so on. Second, the large volume of new facts about the brain, as discussed, are difficult to comprehend, much less assimilate into existing models or generate new theories. (I should add, lest I leave the wrong impression, brain research has clearly contributed to the treatment of many problems, like depression or anxiety. However, there is generally an enormous gap between new research and the ways in which that information is being applied to practical problems.)

When the intern said, "How should I know!?" this project was born. About a year after that incident, I began making daily visits to various university libraries. At first, I read as much as possible on the topic of memory, with a special focus on brain-based theories of memory. Much of what I read focused on basic research with rats or other lab animals, in addition to research on humans, focusing on both normal and abnormal memory conditions. Once I had a general sense of the research, I began reading articles that were more philosophical in nature, but many of them proved to be fascinating and very useful.

The first big payoff of this project occurred after weeks of digging. I discovered tentative answers to my original question, "How might complex emotional memories be stored in the brain?" Although the answer was not universally endorsed by all memory researchers, a possible explanation was nevertheless based on brain research and provided a plausible explanation where none had been available before in many existing counseling theories. (As it turned out, emotional memories are stored separately from "factual" memories, and I will spend a great deal of time in this book looking at different memory phenomena.)

Once I started looking, there was so much to be found. Having hit the jackpot on memory, I then expanded my search to include other topics, such as learning, cognitive development, thinking, and emotion. I counted on the most up-to-date research and writing in all areas. Finding sources was rarely a problem! Digesting it? Well, that was another matter. A few months after starting this project, I had scanned literally thousands of pages of information. Overall, the task was overwhelming.

Quite by accident, I discovered that when I read hundreds of pages, regardless of whether or not different authors agreed with one another, overriding themes became apparent. In the case of memory, the identification of different memory pathways was a critical topic among biologists. In the realm of applied psychology, issues pertaining to the accuracy of memory were paramount in the minds of many. By absorbing enough cutting-edge material, I was able to discern some *general principles* that were extremely helpful, not to mention interesting.

Eventually, I explored over fifty different subject areas. Most articles described brain or behavior research findings, but a small number of general philosophical articles were also reviewed, and they, too, provided insights that were often not available from brain or behavior research alone. Throughout the search, one question remained uppermost: "What general brain-based principles have *practical use and meaning for understanding behavior*?"

Ultimately, twelve *general principles* were distilled from hundreds of sources. In my judgment, these principles are well supported by existing data, and the ideas on which I base any conclusions are highly endorsed by many experts within various fields. I say "many experts" because almost no aspect of brain functioning, at present, is universally supported by all experts within a given field, but I will consider the degree of support that each principle has when the principle is presented.

Here is a synopsis of the general principles: The first two principles concern brain evolution and the social environment in which the brain is thought to have evolved. The third, fourth, and fifth principles address various properties of our brains that clearly distinguish them from computers. Both biological and human factors exert tremendous influence over the ways we "objectively" process data. Principle six focuses on development. Our brains change over time. Principles seven through ten focus on learning and memory. Principle eleven summarizes the many vital functions that emotions serve; and the last principle is about thinking. "Thinking," as a topic, comes last because it is so heavily dependent on all other principles for our understanding.

All principles are compatible with the wave of new brain research. This is an important point because many theories about behavior, as will be illustrated later, are not consistent with new brain research, which results in problems of credibility (not to mention legal problems for practicing therapists).

For the most part, the principles were distilled from many sources. You might wonder why these principles have not been published before since they are based on existing research. Like all contemporary social scientists, I have relied *heavily* on the work of others. Most of the individual ideas in this book, therefore, have been published elsewhere, but as far as I know, no one has brought them together in exactly the way I have done and integrated them into a single theory of behavior. In a similar vein, the

principles distilled here reflect my perspective. If this project were undertaken by an educator or neuropsychologist, other principles likely would
have emerged. What I chose to highlight is clearly a by-product of my
professional training and background. I should add, though, because I am
a trained psychologist, my perspective may seem relevant for most anyone
interested in people and behavior.

Starting with the next chapter, and in each of the following chapters
(except the last one), one or two general principles is presented after a *brief*
discussion of a sample of the data used to distill the principle. The principles are called "Principles for Understanding Behavior" or just "principles." Each one has a unique label. "The Memory Principle," for example,
is presented in chapter 6. Each principle stands alone as a theory in itself,
but when more than one is considered, or when all are integrated together,
it is my hope that a powerful model unfolds that may be useful in looking
at behavior.

After each principle is presented, a section titled, "Implications for
Understanding Behavior" will follow. When we apply psychological theory or even brain research to complex behaviors, we have crossed into the
realm of theory. Thus, the word "implications" is intended to imply that
we are becoming more speculative.

I believe that *all behavior*, whether it is considered deviant or exem-
plary, unfolds in accord with neurological and behavioral principles that
can be studied and understood. At present, though, many behaviors are
clearly not well understood, but each day brings new information that
increases our understanding.

In short, this book attempts to bridge the enormous gap between
neurological and behavior research, with an emphasis on practical understanding and application. In this book, I consider many different behaviors
using simple terms and everyday examples. Many explanations remain
theoretical, at present, but at least those explanations are consistent with
findings from many different scientific disciplines.

This book does not purport to explain the actions of specific individuals. I do not presume to know why, for example, Timothy McVeigh did
what he did. However, we *could* generate some *possible* theories about him
that are consistent with the ideas of this book, if we chose to do so.

Finally, to some extent, this book allows us to look at our own communities because society is comprised of individuals, and by looking at our

past history as a species, and at the way we function at present, we can speculate about future possibilities.

Owing to the jellylike substance inside the skull, *Homo sapiens sapiens*—modern humans—have risen to a position of preeminence among all living plant and animal species, and we hold our position because of our brains. Thus, we might wonder, how secure is our position, and why did we first ascend to this place? This book explores these and many other questions about our very essence.

* * *

Having taken a brief look at what is to come in the pages ahead, let us turn our attention backward a few million years.

which resulted in the technology we hide our wimpy bodies behind

CHAPTER 2

HUMANS IN PERSPECTIVE

It has taken most of human history for the brain to become widely acknowledged as the primary contributor to behavior. At this point, it would be tempting to throw out the past altogether and start afresh with a contemporary look inside the skull. That course of action, however, would be a mistake, in my judgment, because the brain's distant past holds some fascinating clues to understanding our behavior today. Evolutionary theory has produced some extraordinary insights that have provided an invaluable platform for looking at the brain. Thus, it might be helpful to review some of the major tenets of evolutionary theory.

THE THEORY OF EVOLUTION

"The theory of evolution" is somewhat of a misnomer because contemporary scientific views of evolution are a synthesis of many ideas and discoveries.[1] Nevertheless, evolutionary theory is most often attributed to Charles Darwin, who wrote *On the Origin of Species by Means of Natural Selection, or the Preservation of Favoured Races in the Struggle for Life* in 1859.[2] The hallmark of Darwin's theory was "natural selection," which is a concept that continues to play a central role in modern versions of evolutionary theory.

Darwin's ideas about natural selection were founded upon three important observations. First, all living organisms tend to increase their numbers at a higher rate than what would be needed to simply replace individuals who were dying. Second, no single animal or plant completely dominates all environments throughout the world; rather, there is always a struggle for each available niche. Finally, individuals within a group of animals are not identical; thus, Darwin concluded that "natural selection" determined over time which traits and animals would survive in any given environment.[3]

Natural selection embodies the idea that those animals or plants that have the most beneficial traits (adaptations) for a specific environment will live, and those with less beneficial traits will perish. Stated another way, if a trait *increases* an animal's chances of survival, then animals with the beneficial trait have an advantage over animals without the trait.[4]

In the case of two deer in the wild, one with a large lung capacity might easily outdistance a predator. Another deer with smaller lungs might have a quick end to life as a lion's lunch. Having larger lungs, therefore, is an advantage in the wild. In a zoo where deer are not chased by predators, lung capacity is not particularly critical. Adaptations can only be considered in light of specific environments.

Any trait that increases an animal's chances of growing to maturity will also increase the chances that the animal will pass on its genes to offspring. Thus, the heartier members of the species will perpetuate themselves. Conversely, and just as importantly, traits that provide animals with poorer chances of survival are likely to perish. Natural selection is a two-sided coin that involves the selection of the most fit and the elimination of the least fit.[5]

As noted, traits that allow animals to survive through natural selection are called "adaptations," but the word has two connotations. An animal is said to have an adaptation (or positive trait). Gills are adaptations that allow fish to survive under water. Adaptation also refers to a process. A gene mutation—a sudden variation—for example, might result in a whole new line of fish gills that are superior to the old line. Mutation and the advantages passed on to future fish are the *process* of adaptation.[6]

Although modern evolutionary biologists embrace the concept of adaption, it is not assumed that every human behavior or trait reflects an adaptation. There are many examples in nature whereby a characteristic may have evolved for one purpose and later became "co-opted" for another purpose.[7] A good example is the ability to sing an Italian aria. No conceivable prehistoric condition would have demanded the ability to sing an aria, but since many people sing arias, the most likely explanation is that nature "co-opted" one trait and employed it for another purpose. Sounds that evolved for verbal communication or reproductive purposes were modified or "borrowed" for singing. The process whereby a trait is co-opted for a purpose other than its original is called "exaptation."[8]

Furthermore, it is possible for several characteristics to be tied together as a whole package when only one trait was the actual target of natural selection.[9] Height, for example, could have been naturally selected because it provided a survival advantage when looking out over prehistoric savannas, allowing an individual to scan a larger area for dangerous predators. However, stronger legs may have accompanied the added height. Extra height adds weight to an individual's body, so in this example, it is the height—not the stronger legs—that was naturally selected, but the two characteristics often accompany each other.

At this point, we are faced with a potential dilemma. If some traits are the result of natural selection, and others survive only because of their association with naturally selected traits, how can we be certain that a characteristic serves (or served) an adaptive function? This is an easy question to answer. We can never know for sure. However, the inability to know may not matter. An adaptive perspective assumes that most of an animal's behaviors were naturally selected. So when one considers the numerous functions that have collectively been termed "thinking," for example, it is probable that most of those functions were the product of natural selection. On average, therefore, if we view behavior as serving an adaptive purpose, we will be accurate in our assumption most of the time. Since all theories make conjectures that cannot be proven, an adaptive assumption can at least provide a reasonable starting hypothesis when viewing behavior.

Although Charles Darwin expounded the idea of natural selection in his groundbreaking work in 1859, Darwin was completely unaware of the work of another scientist whose findings would become an important part of modern evolutionary theory. In 1866, an Austrian monk, Gregor Johann Mendel, published the findings of his experiments on pea plants, but Mendel's work was not combined with that of Darwin's until the 1930s.[10] It was Mendel who discovered that traits, like eye or plant color, were passed on to offspring by pairs of genes.[11] These findings turned out to be critical for an understanding of evolution because natural selection can work only if animals have a variety of traits. Natural selection implies choice.

A spectrum of possibilities among various traits is called "variation," and variation is guaranteed in several ways, two of which are very important. In humans, when a sperm unites with an egg, each parent donates half the chromosomes necessary for the offspring. The resulting offspring, therefore, ends up with the correct number of chromosomes, but also ends up being different from each parent. By combining different genes in pairs, the number of possible combinations is "scarcely imaginable."[12] The re-sulting child is a whole new model, increasing diversity within the species.

Although recombinations of genes provide an almost unlimited source of variation, perhaps the most critical source of variability is through mutation.[13] Mutations occur when genes are altered. Exposure to radiation, for example, can cause genetic mutations. Most mutations die off, but some survive because they provide a survival advantage for a specific environment.

Very importantly, natural selection can also work on *behaviors* that are learned, *not* just traits that are genetically inherited. Among humans, one reason we can adapt to a wide variety of living conditions is because of learning, and many experts believe that learning is a more efficient means of adaptation than inheriting genetically dictated behaviors. The reason is that learned behaviors are more flexible than actions that are genetically dictated. "For complex animals confronted by constant environmental change, social learning is a more efficient means of acquiring a behavioural repertoire than genetic programming."[14] (As we will see, learning is heavily biased by our genetic heritage, so it, too, is constrained by genetic factors.)

Because some behaviors provide advantages for animals, including humans, whether those behaviors are genetically imparted or learned, there is another possibility that we should consider. Behavior, per se, can result in extinction just as surely as nonadaptive, genetically imparted traits.[15] Just as animals with negative traits can be selected for extinction, animals with learned, but negative, behaviors can also be selected to perish. Bears who learn to raid human trash cans, and therefore forget how to live in the wild, can become extinct because of their learned behavior just as quickly as bears who inherit genetic defects.

As originally postulated by Darwin, evolution was thought to be extremely slow, unfolding over the course of millions of years. However, many experts now concede that evolutionary change *might* occur much more rapidly, and some have postulated that evolution can occur within a few thousand years or even more rapidly.[16]

With this brief overview of evolutionary theory behind us, we can turn to a *short* chronology of major prehistorical events prior to the arrival of humans, just so we can view human evolution in a broader context. The following synopsis highlights but a few milestones during the course of millions of years. Thousands of additional examples could have been included, but for purposes of clarity and brevity, we will consider only a handful.

History of Events

The earth is thought to be between fourteen and fifteen billion years old.[17] Thereafter, all other events that are relevant here occurred more recently.

For perspective, and because we are probably familiar with them, the first dinosaurs are thought to have appeared about *225 million years ago*, during what has been called the Triassic period.[18]

The first mammals and birds appeared approximately *180 million years ago*, during the Jurassic period. It was during this period that dinosaurs ruled the earth.[19] The dinosaurs finally became extinct about *sixty-five million years ago*. Their departure is thought to have cleared the way for a much broader variety of animals, including mammals.

The first primates, from whom we are thought to have descended, emerged between fifty and sixty-five million years ago. Primates are a grouping of animals called an "order," and all members of this group are considered to have the same distant ancestor. In the case of humans, apes, and monkeys, we are thought to have descended from an animal that resembles an Asian tree shrew.[20]

Currently, our closest primate relative is the chimpanzee, based on similarities of genetic makeup.[21] The chimpanzee's distant primate relatives are thought to have emerged some thirty-five million years ago.

Moving rapidly to the present, the first animal who walked upright, *Australopithecus afarensis*, began its tentative walk about four to five million years ago.[22] This group, we should note, was not considered a member of our genus. (*Genus* is the term used to describe a group of species that are similar enough in their characteristics to be related, but *dissimilar* enough that they cannot interbreed. Species are defined as animals that are close enough to breed with one another.)[23]

The Primate Lineage

Long before the first humans, certain traits were well established within the primate lineage, having been imparted through millions of years of evolution. One of the most important primate traits that we share with all other primates is that of being social. We live in groups. Some advantages of group living include better access to food, greater protection, the availability of potential mates, and help with caring for the young.[24] Although we can survive on our own when we have reached maturity, there is no dispute about our social nature.

Although it is easy to assert that we are social creatures, there is a great deal of evidence that attests to this fact. Human infants are completely dependent on older human caretakers for longer periods of time than other primates,[25] human language cannot be learned outside of a social context, and on average, humans who are involved in supportive social relationships live longer than those who live alone.[26]

Human Evolution and Classification

We belong to the genus called *Homo*, which is a group of species that are distinguishable from apes in a number of important respects. Experts believe that each species within our genus shares important traits in common.

First, all *Homo* groups are considered to be a "continuously evolving lineage."[27] (This point is disputed by some experts because the farther we go back in time, the scarcer is the fossil evidence.) Each member of our genus is bipedal, that is, we walk upright on two feet. Walking upright, interestingly, is a trait that did not begin with our genus. Rather, that distinction currently goes to an apelike creature called *Australopithecus afarensis*.[28]

Very importantly, as we will soon see, from the time of the first *Homo* to the emergence of the latest member of our genus, there was a gradual growth in brain size and over the course of a few million years, our brains tripled in size.[29]

The First Tool Users

As already stated, *Australopithecus afarensis* is not considered a member of our genus, so generally we should not be discussing it here, except that it did use stone tools. In other words, the use of tools appeared in nonhuman animals long before humans arrived on the scene. In fact, even chimpanzees use rocks and sticks as tools. What really sets us apart from other animals, relative to the use of tools, is that we have invented highly sophisticated tools, including tools for the purpose of making other tools, whereas tool-using animals must rely on natural objects as simple extensions of their hands or limbs.[30]

Homo habilis

Homo habilis or "handy man" is the first member of our genus, appearing about 2 to 2.5 million years ago.[31] Its brain was much larger than its more primitive, apelike ancestor, *Australopithecus afarensis*. Because of its larger brain, *Homo habilis* may have evolved more sophisticated social customs, which include food sharing, labor division, toolmaking, and the ability to plan ahead and cooperate in a search for food. The theory that *Homo habilis* both cooperated and planned ahead is based on the fact that sources of stone needed for tools have been found some distance from where animals were cut up, suggesting forethought and cooperation.

Because of its tool use, *Homo habilis* has long been classified as a hunter and gatherer. However, they were only about five feet tall and weighed perhaps no more than 100 pounds, and only had the use of very crude tools. Furthermore, fossilized teeth of *Homo habilis* have shown that they had diets similar to those of chimpanzees, which suggests a greater reliance on plant and animal food than would be expected from hunters.[32] Because of these considerations, some now question the likelihood that *Homo habilis* hunted large animals. Rather, it seems more likely that they used their primitive tools to scavenge the remains of animals that were killed by other predators.[33] Because of their relatively large brains, *Homo habilis* may have been able to study and anticipate the habits of larger predators and, therefore, devise efficient scavenging techniques.

One thing is indisputable about *H. habilis*: Not much is known about them, compared with the next *Homo* species. There is very little fossil evidence about *Homo habilis*. To date, their remains have been found only in parts of Africa, so they may have been limited to only a small area. Also, because they are a much older species, fewer fossils have survived.

Homo erectus

Homo erectus is our closest ancestor, presumed to have descended from *Homo habilis*. By comparison to *Homo habilis*, *Homo erectus* was very intelligent, with a brain size about one and a half times larger. To place this size in perspective, *Homo erectus* had a brain about the size of today's average four-year-old human.[34] (When I first read this analogy, I was surprised because *Homo erectus* had weapons, they hunted for a living, and raised families, and it is hard to think of a human four-year-old with that type of responsibility.) As a species, *Homo erectus* lived a span of time that is estimated to be close to two million years. Furthermore, they spread out to most parts of the world. Their level of intelligence, therefore, must have served them well.

In stature, *Homo erectus* was similar to humans, averaging about five feet, six inches in height. Most experts agree that *Homo erectus* was our direct ancestor, emerging about 1.8 million years ago. As mentioned, these individuals were found in many parts of the world, which is a feature that distinguishes them from their earlier cousins, *Homo habilis*.[35]

With *H. erectus*, we begin to get a greater sense of a human-type creature, possibly one with cultural attributes. Their success at traveling throughout the world is testimony to their intelligence and ingenuity. Their wide travels suggest that they planned ahead and cooperated with one another. Certainly, *Homo erectus* developed more advanced stone tools, and they were likely to have been the first to fashion tools for the purpose of making other tools.[36] Their diets, too, were more similar to ours today. Clearly, *Homo erectus* was a true hunter and went beyond scavenging.[37] Finally, they are credited with knowing how to start and use fires.[38]

In spite of their advances, *Homo erectus* remains an enigma. They left no evidence of rituals: no figurines or statues, no signs of art, no evidence of burying their dead. Even their tools were invariably simple and utilitarian.[39] They *may* have had spoken, symbolic language, but this point is debated by experts. In short, beyond a few notable and significant advances, *Homo erectus* never achieved even a fraction of the accomplishments attributed to their successors.

It is widely believed that *Homo erectus* evolved into contemporary humans, although a remote possibility remains that modern humans replaced *Homo erectus* (instead of evolving from them). In any event, the transition to the next human species, *Homo sapiens*, is hypothesized to have begun perhaps five hundred thousand years ago.[40] However, the earliest *Homo sapiens* did not have the same appearance as humans do today, even though their brain cavity (and presumed intelligence) was gradually approaching modern standards.

Homo sapiens

It has long been thought that *Homo sapiens* evolved in Africa and then moved to other parts of the world. Now, a new debate rages on this issue.

Some experts believe that *Homo sapiens* may have evolved in different regions in the world at roughly the same time.[41] Part of the disagreement is fueled by the lack of fossil evidence and uncertainty about how to classify some remains. In all populations, there are naturally occurring variations from one individual to the next, so it is difficult to reach an agreement on how to classify all fossils.

"Modern humans" (*Homo sapiens sapiens*) are those people who look like and have all the attributes of people today. The first **completely** modern people appeared about fifty thousand years ago, or even longer.[42] However, humans that were genetically identical to us are much older, first appearing from one hundred thousand to two hundred thousand years ago.[43]

THE RISE OF CULTURE

About forty thousand years ago, humans began to reach many of the milestones that we typically associate with being human. From forty thousand to ten thousand years ago, during a period called the Upper Paleolithic period, humans were really setting themselves apart from all previous human and animal groups. During that time, bows, arrows, spear throwers, and tiny, replaceable blades were invented[44]; and recent discoveries in France date the first paintings to about thirty thousand years ago.[45]

The first humans, and their *Homo erectus* and *Homo habilis* predecessors, lived in bands. (Information about prehistoric bands is often generalized from contemporary hunters and gatherers, but some information can be logically inferred through fossil finds.) Bands were organized loosely, probably consisting of groups of people who were related to one another by blood or whatever passed as marriage in those days. Further, bands were believed to have been egalitarian for several reasons.[46] First, within hunting and gathering societies, the bands were small, so all members were important to the band's survival. The individuals within the group needed to cooperate with one another for the survival of the entire group. Also, in contrast to people who came later, hunters and gatherers had few personal possessions because property would have been a burden to those who were constantly on the move.

All band sizes were estimated to be relatively small, averaging from twenty-five to sixty individuals, depending strictly on the availability of food.[47] Typical work groups, therefore, like hunting parties, would have been even smaller. These numbers are important because they reveal something about the social conditions in which the brain evolved.

By forty-five thousand years ago, humans were widespread through-out most parts of the world. As human population numbers increased, social structure may have also begun to show some changes. Specifically, in areas where food was most abundant, several bands may have come together briefly for part of the year. This hypothesis is supported by the fact that in some areas, greater numbers of art objects and other artifacts have been found together. However, several bands could not have re-mained together long because food supplies would not have supported large groups prior to the domestication of crops and animals.[48]

THE RISE OF AGRICULTURE AND ANIMAL DOMESTICATION

Prior to the rise of agriculture, some hunting and gathering societies may have experimented with lifestyles that were somewhat less mobile, especially in regions where food was very plentiful.[49] Overall, though, during the Upper Paleolithic period people were hunters and gatherers, relying mostly on large, migratory herds of animals, like antelope, bison, mammoths, and elk.[50]

End of an Era

The transition from hunting and gathering to animal and crop domes-tication literally changed the face of the earth, but perhaps the face of the earth had already changed. The Ice Age was drawing to an end.

From 15,000 to 10,500 B.C., the oceans rose about seventy feet. By 7,500 B.C., the seas had risen another two hundred feet. All of this changed the face of the planet, especially in Scandinavia and Great Britain. In some parts of the world, enormous land masses were being exposed by receding ice sheets, but in other areas, water from melting ice was inundating coastal areas.[51]

In the face of massive geographical changes, food sources were also affected. By 10,000 B.C., many large game animals had become extinct. Although many game animals did survive, their numbers had decreased to the point where humans had to change their ways. For most people, that meant learning other methods of obtaining food.[52] Although climate changes destroyed some food sources, humans were fortunate because rising oceans provided a new source of food. Humans learned to exploit the bounty of the seas, relying more heavily on fish and shellfish for food.

Regardless of the reason, humans *did* shift from hunting and gathering to agriculture, and that monumental change was completed approximately ten thousand years ago. Furthermore, there is little doubt among experts that the change to agriculture led directly to the birth of cities.[53]

When humans ceased to be hunters and gatherers and began to cultivate the land, there were enormous changes in social structure. Once humans began to live in permanent settlements, population numbers began to rise, accompanied by greater exploitation of local food resources and the adoption of trading practices over larger distances. Food-producing societies allowed for private ownership, which in turn created class differences that had never existed before. As a result, some individuals became wealthy because they controlled the local resources, but most did not.[54] In short, once permanent settlements were established, new social conditions emerged.

Perhaps the most important change in social structure involved a change in the size of social groups. In contrast to hunting and gathering bands, ancient settlements ranged in population from a few hundred inhabitants to as many as a million. With increases in population, as we would guess, the complexity of society increased. Governments arose and became centralized. Taxation was invented. Greater complexity required a means of keeping records, so written records were eventually invented. With centralized governments, nonfood-producing jobs also emerged for the first time, to include such professionals as the clergy, civil servants, and professional soldiers. A broad range of craft specialists also sprang up, such as metalworkers, potters, and so on.

When reviewing our distant origins, there are many details to be digested, and a good many theories to be considered. It might seem as if we cannot draw any useful conclusions from the past. A recent article in *Time* magazine underscores what appears to be disarray in the field of anthropology: "Findings announced in the past two weeks are rattling the foundations of anthropology."[55] This quote implies that the field of anthropology has been turned upside down, but that is not the case for our purposes. On the contrary, when we review the data, as we have done here in a cursory manner, important conclusions *can* be drawn. If anything, it is encouraging to see how much substantive *agreement* there is on some critical issues. What is described as controversial in other fields is not an impediment for our study here. Two important principles relating to the brain and behavior can be safely extracted from the information gleaned:

PRINCIPLE 1: THE BRAIN GROWTH PRINCIPLE

The brain of modern humans reached its large size long before humans began to grow really inventive and create such things as written language, art, agriculture, steel tools, and large cities.

When written, this principle appears painfully obvious. However, it is important because it helps to provide perspective.

Prior to the growth spurt of the brain, our human ancestors had made some advances, but not many, in comparison to the changes that came *after* the brain's enlargement. This is important because it supports a hypothesis that the modern brain enabled us to become more inventive.

PRINCIPLE 2: THE SOCIAL ENVIRONMENT PRINCIPLE

The brain's most recent growth or evolutionary change occurred during a time when our ancestors lived in small groups or bands of twenty-five to sixty individuals.

Evolutionary theorists are in disagreement about many aspects of our distant origins, including such important matters as whether we evolved gradually or more suddenly. In either case, there is little disagreement that our brain reached its current size perhaps two hundred thousand years ago.

From the time when the new, larger brain first appeared, all humans lived as hunters and gatherers, with no known exceptions, until about ten thousand years ago. This suggests that our brains evolved under social conditions that no longer exist today for most of us.

IMPLICATIONS FOR UNDERSTANDING HUMAN BEHAVIOR

Much of the foregoing discussion focused on ideas that are well regarded by some experts, but ideas that are not without criticism. Because human evolution deals with subject matter that cannot be definitively proven, it is most useful perhaps to reflect on what is believed about our prehistoric roots and continuously ask ourselves, "How would this behavior or trait today have served an adaptive purpose under the conditions when our prehistoric ancestors were alive?" As we will see, humans have a strong, innate tendency to stereotype, and we will later consider what adaptive purpose that stereotyping might have served.

Compared with known changes that occurred during human prehistory, the pace of modern change is very rapid. The technological era has brought about changes in the last forty years that were unfathomable even fifty years ago. This is an important point because it raises the issue of how quickly we can adapt to new changes, particularly since there are no indications that our brains have changed from when they first appeared in their current size, possibly two hundred thousand years ago. Size continues to be a critical issue, because the ratio of brain size to body size is one of the best indicators we have of intelligence. Since we cannot measure the intelligence of our deceased ancestors, we have to resort to skull sizes and inferences about behavior to make educated guesses about intelligence.

Although our brains have allowed us to make unprecedented changes to the world, we can conclude that humans do not fully understand the long-range consequences of those changes, partly because many long-range consequences have yet to occur. At present, we can make major alterations to our social and physical environments so rapidly that we may be unable to measure the impact of those changes until long after they have occurred. For example, televisions were a standard feature in most American homes long before anyone knew about the effects of television viewing on the growing brain; and even now, experts are not in agreement about all the possible ramifications of television viewing.

Certainly, most of us would agree that humans are capable of making unprecedented modifications to our physical and social environments, compared with our *Homo erectus* ancestors or their primate cousins. At the same time, though, we humans are among the most adaptable of any known mammal species. This latter assertion is based, in part, on the fact that humans have learned to survive in all parts of the world.

If modern humans are living in a vastly altered world, some important questions arise, given the intimate relationship between health and environment. As humans exert greater influence over their environments, could we collectively be creating changes that are health- or life-threatening? At the national level, some experts believe this is the case.

Bruce Perry is a renowned neuropsychiatrist with the Department of Psychiatry and Behavioral Sciences at the Baylor College of Medicine in Houston, Texas. He has asserted that over five million American children and adults are exposed each year to social conditions, such as violence, that have devastating effects on personal functioning and intellectual growth. In his words, "The profound impact of domestic violence, community violence, physical and sexual abuse and other forms of predatory or impulsive assault cannot be overestimated."[56]

It is certainly tempting to draw an association between major shifts in social structure and recent, now infamous, acts like those of Theodore Kaczynski, the Unabomber, or Timothy McVeigh, who bombed the federal building in Oklahoma City. However, the average person does not become a mass murderer, nor appear to manifest problems serious enough to warrant a forced intervention from either law enforcement or mental health authorities.

There are, however, clear effects of the change from hunting and gathering to our contemporary lives. We have seen a tremendous *increase in the diversity* of human lifestyles and social groups. What this diversity might mean, though, is difficult to assess without further information about how the brain functions. As we learned earlier, the concept of adaptation is only applicable to a specific environment. Thus, it is *very likely*, if not certain, that some humans are well matched to their environments, and others are not. Beyond this general conclusion, it is not possible to make specific inferences about "the environment" and behavior. Rather, we must specify which environment we are talking about relative to specific individuals.

CONCLUSION

Regardless of whether we think of evolution as exceedingly slow or as occurring more rapidly, regardless of whether we look at our most distant human predecessors (*Homo habilis*) or at our most recent ancestors, one important point remains unchanged: For most of human history, humans lived in small social groups, averaging from twenty-five to sixty individuals, surviving through some combination of scavenging, gathering, and hunting.

Furthermore, *all* of the most important inventions that we associate with humans—including modern technology, the rise of agriculture, and the concomitant rise of cities; the invention of sophisticated tools, written language, art—all occurred *after* our brains reached their current size. However, humans did not immediately begin to make major alterations to their social environments when the first large brains emerged. Humans continued to live as hunters and gatherers for thousands of years after the large brain emerged.

Other generalizations also appear warranted at present. First, modern societies are quite different from hunting and gathering societies in terms of numbers of inhabitants, yet today's brain evolved within small hunting and gathering communities with no known exceptions.

Further, modern civilization has resulted in many changes to our social structure, but those changes have not influenced all people evenly. Some children live in poverty and neglect, others live in affluent, caring families. This fact is important because, in contrast to our human ancestors whose children were subjected to the same or similar social structures, today's world provides children with enormously diverse living conditions, which means that some children get all they need to develop and thrive, and others do not.

Finally, humans are enormously adaptive, so the impact of various social changes cannot be easily assessed without looking at specific individuals within the confines of specific social and physical environments.

This chapter produced the first two of twelve "Principles of Behavior," restated here:

PRINCIPLE 1: THE BRAIN GROWTH PRINCIPLE

The brain of modern humans reached its large size long before humans began to grow really inventive and create such things as written language, art, agriculture, steel tools, and large cities.

PRINCIPLE 2: THE SOCIAL ENVIRONMENT PRINCIPLE

The brain's most recent growth or evolutionary change occurred during a time when our ancestors lived in small groups or bands of twenty-five to sixty individuals.

* * *

Now that we have considered ourselves from a prehistorical perspective, we can look more closely at the big brain that has resulted in so many changes.

CHAPTER 3   THE FIRST NETWORK

Climates changed and some forests receded. Driven by hunger, some of our distant ancestors began to look downward from the safety of their trees. At first, a few left their arboreal homes forever, but their past lives in the trees served them well. Carrying the first rudimentary tools fashioned from tree limbs, some of our distant relatives used sticks to scratch for food or drive off predators. With sticks in hand, some learned to walk upright. Just as their earlier cousins had relinquished their homes in the trees forever, a small number of primates relinquished their pasts and walked upright from then on. That was four to five million years ago.

During the ensuing five million years, the brains of some primates grew from the size found in chimpanzees, almost tripling in size to that found in "modern" humans.[1] That growth, however, did not impact all parts of the brain equally. Rather, the most significant changes, the greatest enlargement, occurred in the "new cortex," also called the neocortex. This and other sections of the brain will be explored shortly.

TERMINOLOGY

On the whole, our brains contain an estimated eighty billion information processing cells called *neurons*.[2] Other scientists have not been so stingy with their estimate, placing the number closer to one trillion.[3] Different scientists provide varying estimates of the number of neurons, probably because the quantities are so vast that only estimates can be made. Regardless of whose count we choose, brain cell numbers are enormous, and even the smallest functional unit of the brain may contain literally millions of brain cells.

31

Limbic system = memory emotions, learning (handwritten annotation)

In spite of crowded conditions within the skull, the brain has four ventricles or reservoirs that are filled with cerebrospinal fluid. This fluid provides nutrients, removes waste products, and provides a protective bath.[4] Without protective fluid and the bony encasement of the skull, our brains could not survive. With the consistency of jelly, the weight of our brains would destroy millions of brain cells in an unprotected environment.[5]

Unfortunately (very unfortunately from my point of view), not all scientists describe the brain in exactly the same way, nor do they use the same terminology. Rather, there are many brain parts that have been given different labels. The cortex is a good example. It is referred to as the cerebral cortex, the cortex, or the neocortex.[6] Problems with terminology should not be a problem in this study, however, because this book is focused only on general brain principles that relate to behavior. It is not a book intended to describe specific brain structures in detail. Thus, I only refer to a handful of major regions of the brain, unless more details are necessary, in which case additional information will be given. There are, however, some expressions we need to be clear about, some of which are rather nonscientific.

Clusters, Nuclei, Parts, and Structures

Within the brain there are many groups of cells that cluster together and form recognizable shapes. A group of cells that collectively form an identifiable part is called a nucleus (from the Latin term for nut).[7] The plural of nucleus is nuclei, but I also use the terms "parts" or "structure" for a little variety. Many brain structures are actually composed of several nuclei. Nevertheless, all of these terms refer to groups of cells within the brain that can be recognized (by experts) by their shape and location.

System

The term "system" (or systems) does not refer to a specific part or fixed structure within the brain. Rather, a system refers to collections of nuclei or cells that perform a specific *function*. Emotions, for example, are usually attributed to the limbic system. In turn, the limbic system consists of many identifiable nuclei, and different emotional reactions may depend on different parts of the limbic system.[8] Specifically, anger and fear are both considered to be emotional reactions, but each may depend on different areas of the limbic system or other parts of the brain. Just as important, neither depends on the exact same group of brain cells as the other.

Function

The term "function" does not refer to a tangible or physical part of the brain. It obviously describes something the brain does. "Thinking" is one function of the brain, but thinking is an imprecise term that may encompass such phenomena as conscious thought, problem solving, contemplation, imagination, or making a decision about where to go on vacation.

Region

The brain can be divided up or mapped into various areas or regions. Just like the Eastern United States can be thought of as a region, so, too, the brain can be divided up into broad regions. Each region contributes to many different functions. The frontal region of the brain, for instance, is a large area that contributes to such phenomena as judgment, forethought, and goal-directed behavior.[9]

The brain is very crowded and complex. Fortunately for brain scientists, just as we can walk through the mountains with a map that designates distinctive landmarks, the brain can also be "navigated" on the basis of distinguishable landmarks (in some areas). Nuclei, with their distinctive shapes, aid in the mapping of the brain, and in other areas there are deep fissures and noticeable ridges that also help to orient researchers to specific locations. Finally, scientists have developed maps of the brain that are divided into numbered coordinates.[10]

Biologists use terms like *caudal* (toward the tail) and *medial* (toward the midline) when discussing direction within the brain.[11] Although these terms are very useful, they require a great deal of study before a person could become proficient in their use. Therefore, these labels will not be used in this study because they are not needed for the points we will be considering. The rather unscientific terms of "lower" and "higher" will be used because they are easy to understand, they are often used by laypersons, and they suffice to convey the points we need to consider.

When used in reference to the brain, "lower" and "higher" have two connotations. First, the terms are used figuratively. Higher functions often denote those processes that are more recent in evolutionary origin, like speech fluency or the ability to logically solve problems. The term implies more sophistication, such as in higher level mental processes.

Higher can also be used in its literal sense, but only if we are standing in a rather awkward position. Since our skulls are very crowded, our brains are curved. The top of the brain refers to what would be the top if our brains had grown in a straight line (which they did not). The top is toward the front of our skulls, behind the forehead. Thus, only if we were standing upright and staring toward the ceiling is the top of the brain actually on top.

Figure 1 depicts a human brain that is sliced approximately down the middle of one cerebral hemisphere, from front to back. It will be useful to refer to figure 1 as you consider the following overview of the human central nervous systems (CNS).[12]

CENTRAL NERVOUS SYSTEM

The central nervous system comprises the brain and spinal cord (or spinal column), so we will begin our discussion with a look at the spinal cord.[13]

The Spinal Cord

Our spinal cords are important for this discussion because they help place our brains into a broader, evolutionary context. The spinal column is

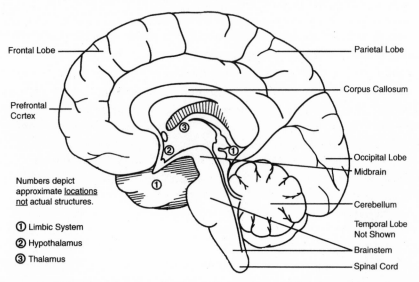

Figure 1. Side view, front to back, of human brain. Only one hemisphere is shown. Temporal lobe is not depicted because it covers most of the lower structures shown. (Courtesy of "NIDA Notes," 11, U.S. Dept. of Health and Human Services, National Institute of Health, Sept.–Oct. 1996, p. 19.)

an extremely ancient type of nervous system, having its origins millions of years ago. It is a segmented tube that extends the length of an animal's body.[14] In animals like worms, each segment receives and sends messages from sense receptors located on a section of the body adjacent to a tube segment. In worms, spinal cords operate in a "stimulus-response" fashion. An incoming signal from the worm's body triggers an *automatic* response—no thinking there.

Within humans, the spinal cord has many vital functions. It carries "messages" (nerve impulses) from the brain to the body and from the body to the brain; and it also controls certain reflexes. When the doctor taps your knee with a hammer and your leg kicks out, that is a spinal reflex.[15]

The Brain

As a whole, our brains are symmetrical organs, weighing about three pounds and consisting of two hemispheres or two sides.[16] Most of us have probably heard the term "split brain," which refers to the fact that our brains are composed of two hemispheres. We will address the significance of this split later.

Generally, the brain is described as having three major divisions, from bottom to top. In other words, the brain, as a whole, has three major areas from the bottom up. But the brain is split into two sides, so each side or hemisphere contains the three major sections, also from the bottom up. Each hemisphere contains the three divisions we are about to consider.

The three major parts of the brain are the *forebrain*, the *midbrain*, and the *hindbrain*.[17] Each of these sections can be further divided into many additional sections, but we will concern ourselves with only a small number of brain areas. (It should be noted that not all science writers describe each section of the brain in exactly the same way. However, the following details are generally accepted.)

The Hindbrain. The most prominent structure in the hindbrain is the *cerebellum*, which can be seen in figure 1. The cerebellum is connected to the spinal cord in the lower back of the skull.[18] The cerebellum is involved with coordinated movement, like standing and walking. Importantly, the cerebellum grows quickly after birth, almost reaching its full size by the age of two.[19] This is important because we need our cerebellums in good working order to learn to walk and stand. Finally, the cerebellum also supports certain types of learning.

The Midbrain.   In humans, the spinal cord extends upward into the middle of the brain and becomes wider. The point where the spinal column becomes wider (inside the skull) is called the *brainstem*. As you can see in figure 1, the top area of the brainstem is called the midbrain.

As we would expect, since the brainstem is the extension of the spinal column as the column enters the brain, the brainstem controls vital functions like breathing, blood pressure, pulse rate, and both sleep and wakefulness. Also, sensory information coming and going from the spinal cord also passes through the brainstem. Finally, for understanding behavior, a rudimentary form of learning occurs within the brainstem.

The Forebrain.   The forebrain refers to a general area of the brain, just as Northeast refers to a general area of the United States. As the complexity of mammals evolved, the brainstem became much more sophisticated, resulting in the forebrain that contains *several* areas of particular interest.

The *limbic system* is part of the forebrain (its general location can be seen in figure 1). Generally, the limbic system is involved with memory, learning, and emotion.[20] The limbic system is an ancient system, but not as old as the brainstem. At one time, the limbic system was the crown jewel of evolutionary achievement. Rather than being forced to respond automatically in a stimulus-response fashion, the limbic system provided animals with a choice, for example, an enraged dog could either run or attack.

Residing close to the general area of the limbic system, we encounter a tiny little structure called the *hypothalamus*. (The general area of the hypothalamus is indicated in figure 1.) Comprising a mere .3 percent of the brain's weight, it is essential for regulating body metabolism, food and water intake, sleep, and temperature regulation, to name a few of its jobs.[21]

Above the hypothalamus is the *thalamus*. This is the brain's primary relay station for both sensory data and information about movement. The thalamus directs signals from lower parts of the brain to the correct locations higher in the brain.[22]

The largest part of the brain are the two cerebral hemispheres, which are covered by a substance that looks like tree bark, hence its name, *cortex*. The human cortex is the newest evolutionary brain innovation, which is why it is also called the "new cortex" or "neocortex." It is the part of the brain that is believed to have grown the most during the last five million years of brain evolution. (Figure 1 shows much of the cortex; the lower part is missing to expose the inner workings of the brain.)

The cortex is shaped like the letter C, with the opening facing downward; therefore, the structures immediately underneath the C-shaped cortex are also forced into a C-type configuration (facing downward). The cortex covering each hemisphere is further divided into regions or lobes, and lobes are named after the skull bones closest to each lobe, thus they are the frontal, parietal, temporal, and occipital lobes.[23] Our frontal lobe helps us plan ahead, learn, and support "working memory." It also supports emotional functioning in conjunction with the limbic system. The parietal lobes receive sensory information from the eyes, ears, nose, and tongue and help to integrate that information. At the back of the head are the occipital lobes, which are involved with vision. Finally, the temporal lobes are heavily involved in language. (The temporal lobe is not depicted in figure 1. If it were, it would cover the midbrain, most of the brainstem, and the area of the limbic system. Also, each brain lobe constitutes a rather large area of the brain, so the foregoing discussion was grossly oversimplified.)

The neocortex is an evolutionary adaptation, reaching its pinnacle in humans. The human cortex is commonly thought to provide us with many of the characteristics that we consider uniquely human, like the capacity to reason, the ability to use symbolic language, plan ahead, write symphonies, or even uncover the laws of physics.

As we have seen, the brain is often segmented into parts. It can be divided from the "top" down, into the forebrain, midbrain, and hindbrain; or it can be divided from side to side, depicted as the "split brain." Because of these naturally occurring demarcations, some behavioral scientists have developed theories that capitalize on these naturally occurring separations within the brain. "Top down" theories are those that emphasize the differential pattern of brain evolution. One such top down theory refers to the triune brain, consisting of a reptilian brain (corresponding to the brainstem), the visceral brain (corresponding to the limbic system), and the neocortex (the new cortex).[24] This type of division has been useful in conceptualizing some psychiatric problems, such as anxiety disorders.

As noted, the term split brain has also been applied to the brain. Both hemispheres are connected by a bundle of nerve fibers called the *corpus callosum*. There is no doubt that the right hemisphere receives input from the left side of the body, and vice versa, but in addition, some functions *appear* to stem predominantly from the right or left brain, as a general rule. For example, facial recognition is often thought to be dependent on the right hemisphere.[25]

Clearly, dividing the brain into sections, or dividing it from top to bottom, has practical utility, *but a blaring note of caution must be sounded.*

The concept of a triune (top to bottom) or a split brain can be *mislead-*

ing. Complex behaviors, like talking on the phone, keeping notes, and tapping a pencil in frustration, often draw upon several areas of the brain simultaneously, cutting across major regions, hemispheres, and evolutionary boundaries.[26] Thus, when a single behavior is attributed to the "emotional mind," such as throwing a dish in a fit of anger, we may be left with the impression that a behavior was *caused* by a specific area of the brain, but such a conclusion is not invariably accurate. In general, many areas of the brain almost always contribute to complex behaviors.

More specifically, in a task that might seem simple to you or me, such as talking about a topic that is well known, brain imaging studies have shown that many areas of the brain are activated. Even when different individuals are performing the same task, different areas of the brain may be called into play from one person to the next. Further, the same task might activate the same brain regions from one person to the next, but to a greater or lesser degree.[27]

Importantly, brain areas with older evolutionary origins do not necessarily function in humans as they did (or do) in nonhuman animals, so top down models often attribute functions to the human brain that may be inaccurate. In rodents, there is a structure within the limbic system that is highly attuned to odors. In humans the same structure shows almost no sensitivity to smell.[28] In short, the area of the brain labeled the "limbic mind" demonstrates significant differences in humans when compared with the limbic systems of rats.

Having sounded the foregoing warning, we must now try to paddle back upstream. *With little doubt, different parts of the brain did emerge during different periods of evolutionary history, and this principle has some utility for understanding behavior and mental processes.*

As I sit here working on my computer, my body posture is being controlled automatically by various muscles within my body. Those muscles, in turn, are monitored completely outside of my awareness, by areas within the brain that evolved long before the neocortex. The distinction between different brain regions with different evolutionary origins *can* be important. The way we would successfully modify a learned fear is different from how we would modify an incorrectly learned history fact. As a general principle, the notion of varying evolutionary origins, or the idea of a split brain, *can* be useful, but the principle must be applied in a most judicious way.

Up to this point, I may have inadvertently given the impression that different brain parts "cause" specific behaviors. *That conclusion is not accurate.* Rather, different nuclei, or cells within a single nucleus, often work in concert with other cells to accomplish the task at hand. Even a single perception, like the recognition of a rose, *may* require neurons located in several areas of the brain.

In spite of new imaging techniques, much of what is known about the brain comes from studies of animals or people who have sustained brain damage. Very often, when the brain is injured, there are typical symptoms, depending on the area of damage. However, a study of injured brains cannot reveal *all* the areas that *may* contribute to a specific behavior. An analogy might make this point more clearly than I am able to do through a written explanation.

Any number of problems can disable a car. We can disconnect the battery, empty the gas tank, or pull out a bunch of wires under the hood. In none of these instances, though, would it be accurate to conclude that a single factor is responsible for those times when the automobile functions normally. The fuel, the spark plugs, the battery, and hundreds of other parts, collectively, must function properly for the automobile to perform. This analogy holds true for the brain.

Many brain structures (or parts of larger structures) are *necessary* for some behaviors, but rarely is a single part of the brain the cause of a single behavior (or mental process). There are some exceptions to this rule, but it takes highly trained experts in fields such as neurology or neuropsychology to determine where specific problems relating to brain functioning and behavior *might* be localized within the brain (if, *indeed*, the problem can be localized at all).

Consider "memory" for another illustration. Our brains house many types of memories. Some types are found throughout the brain; other types are restricted to small regions within the brain, largely kept separate from other brain areas and functions.

The growing consensus about the brain's overall functioning is one in which small clusters of neurons, with highly specialized functions, are *temporarily* linked with other such specialty groups. Linkage, in turn, is performed by "association" and "executive" regions of the brain.

An association area combines different types of information. When we see a beach ball, properties of color, size, and shape have been integrated *before* we consciously recognize the object as a ball. Our brains actually contain areas that have been labeled with the term "association," such as the visual association cortex.[29] To illustrate how this area works, consider the case of a man who sustained damage to the visual association cortex. Afterward, he could no longer name or recognize objects by sight, yet he is *not* technically blind because his eyes are perfectly normal and still register incoming light. His brain, however, cannot make sense of that data.

"Executive" regions take integrated data, such as the conclusion that different colors combined with a specific shape is a beach ball, and decide on an appropriate action—either catch the ball, hit it, ignore it, or run with it.

One of the greatest difficulties in understanding the brain is that all neurons and structures within the brain have a purpose or function. Thus, when a brain structure is damaged or destroyed, typical symptoms occur. In the foregoing example, we learned that damage to the visual association area can result in the inability to name objects, although the individual is not blind. One reason the brain poses difficulty for those who study it is because even simple-appearing behaviors may require millions of neurons and different brain structures. Thus, what may seem like a simple behavior might require widely dispersed cells within the brain.

Neuroscientists are only now discovering what combinations of cells are essential for what tasks—and to date, only the simplest of tasks have been studied, because research technology has not progressed to the point that highly complex behaviors can be studied (such as what brain regions are involved when a young woman falls in love with her future husband).

To understand how the brain works, another analogy can be helpful. Our brains are roughly analogous to a corporation. Many employees can be relied upon from different parts of the company for a large job. Several workers are under the supervision of a midlevel manager. The entire job, though, is overseen by an upper-level executive who keeps track of the big picture. With regard to this analogy a caveat is needed.

When neural pathways are drawn upon constantly, patterns may develop into habits. Whether it is a physical habit, like an ability to throw a ball; a mental pattern, like the propensity to fret and worry; or an emotional habit, like throwing a cup in a fit of anger, the more times any behavior or mental process is repeated, the more likely it is to reappear. The current presumption is that relatively stable neural pathways form.

In sum, even mental functions that appear routine may draw upon widely dispersed areas within the brain. As an example, if you are on the phone arguing with a mechanic about your car, taking notes on the conversation, and trying to keep your eye on a pot of simmering stew, these activities would require input from all over the brain.

Aside from its tendency to utilize cells from widely dispersed areas to accomplish a task, the brain has another tendency that has often confounded educators, psychologists, and researchers alike. The brain often segregates information that seems like it *should not* be segregated.

"Segregation" is a theoretical concept, not a biological one. The term is not indexed in neuropsychology or biology textbooks. However, the concept of segregation is based on well-documented observations of behavior. To illustrate, I could be deathly afraid of spiders (even spiders that are harmless to people). I could also spend several hours learning about spiders in biology class and *know* that most varieties are completely harmless to people. Nevertheless, if confronted by a harmless spider at home, I might jump out of my skin. The knowledge about spiders gained from biology class, in and of itself, does not automatically diminish my spider fear. The "spider fear" (a phobia) is segregated and stored separately in the brain from "knowledge about spiders." *The brain is a master at this type of segregation.* At present, it is not known how this segregation phenomenon is maintained, but many writers have provided speculative comments that might be enlightening.

According to one researcher, if we consider the number of connections made between nerve cells within the brain, each brain cell can communicate with only a small fraction of the total number of brain cells. Thus, as a network, the brain is "vastly underconnected."[30]

Other biologists describe a much larger number of connections, but they also conclude that most connections between brain cells are typically made with cells that reside nearby. In other words, even though the brain may have a vast network of connections, most connections are between cells that abide close together. This description of the brain would still leave open the possibility of underconnectedness.

As we saw, some scientists have written about the different evolutionary origins of different brain structures, with the conclusion that brain structures with different birthdates often control different types of functions. This, too, is a type of segregation. Referring to the spider example from before, the "knowledge of spiders" and "fear of spiders" are likely to be maintained in separate areas of the brain that evolved during different periods of evolutionary history. The segregation of information is certainly consistent with known attributes of the brain and consistent with many observed behaviors.

Regardless of how the brain accomplishes segregation, observations and studies of behavior suggest that information learned in one context is *often* unavailable for use in another situation, even when the other situation would be an ideal place for the stored knowledge.[31] For example, I could spend all day in physics class learning about the properties of a fulcrum. On the way home, though, my car might get stuck in the snow, and the idea of a fulcrum might not enter my mind at the exact time when knowledge about leverage might do the most good.

Knowledge and information appear to be stored *contextually*; that is, our brains are highly attuned to context. Information gained in one setting might readily be called upon in that setting again, and the same information might remain dormant outside the context where the information was originally learned. This is a broad generalization, but we will revisit this idea in subsequent chapters.

The foregoing overview was broad and sweeping, of necessity. In all likelihood, it will take neuroscientists many years to penetrate the remaining mysteries of the brain, especially when complex thinking functions are concerned. (This assumes, of course, that the brain's remaining mysteries can be penetrated.)

In the meantime, though, there is a general principle that can be distilled from what has been presented—a general concept that has enormous utility for thinking about behavior and generating possible explanations for many common occurrences.

PRINCIPLE 3: THE NETWORKING PRINCIPLE

Our brains are highly complex information-processing systems or networks that often segregate information. The concept of a "network" has two implications. First, for most behaviors and mental activities, such as thinking, many parts of our brains are temporarily linked together by brain areas called "association" and "executive" regions. Second, our brains are made up of many parts, some of which evolved during different periods of evolutionary history.

To illustrate "networking" in action, consider what you are doing now. As you silently read the words of this book, hundreds of thousands of neurons are required for the task. If you were to listen to the same words being read out loud, other cells would be brought to bear on the task. If you were to speak the words out loud, and add animated arm gestures, like an actor on a stage, still other parts of the brain would be brought on-line.

As we go about our daily business, the brain orchestrates various activities from within the skull, manifesting a *constantly shifting pattern of cells or nuclei to accomplish our goals.* As a general rule, any change in perception, experience, thought, emotion, movement, no matter how minute, is supported by a slight or major shift in the exact configuration of brain cells required for that activity.

The networking principle also contains the concept of segregation, as discussed. Our brains constantly perform many functions simultaneously, and some functions are carried out relatively autonomously from others. As I think about this sentence, my cortex is involved, but part of my brainstem is regulating my heartbeat. The regulation of heart beat and "higher-order" thought processes originate within different areas of the brain.

The ways in which I have described the brain are based on emerging brain research. The actual terms used, though, like "segregation," are conceptual because, at present, many functions of the brain cannot be precisely defined in biological or neurological terms. The experts would love to talk about thinking by identifying the exact nerve cells involved, along with an understanding of how those cells solve complicated physics problems. Unfortunately, there is a big gap between observed behaviors (solving a physics problem) and what can be directly linked to known brain processes because current brain-imaging techniques are far too imprecise to track highly complex activities.

Although the foregoing and following discussions are theoretical in many respects, as we learned previously, all of us use theories to explain and understand behavior. Nevertheless, as theories go, the ones in this book are consistent with new research on the brain and behavior. We can now put our theory to some practical use by looking at some common behaviors.

IMPLICATIONS FOR UNDERSTANDING BEHAVIOR

As the brain evolved over millions of years, many "older" functions and capabilities were maintained, and periodically, newer capabilities were added to the mix, presumably because of environmental influences and natural selection. Thus, the brain is, in effect, many brains in one. *Within our heads are many systems that produce learning and memory. There are different types of thinking, and a multitude of networks contribute to motivation.*

Unfortunately, very different systems often share the same label, like "learning." Two learning systems, though, as illustrated previously in the spider example, can be quite different in terms of the exact combination of neurons or nuclei that support each function. From a practical standpoint, this principle is important because if we want to influence learning, we need to know something about the system we wish to modify.

Moving beyond the spider example, consider the problem of teaching a three-year-old child to stay out of the street. If we raise our voices and shout "no," just as he or she steps off the curb, the child may not repeat that misbehavior again. If, however, we say "no" after he or she has been in the street for a while, the "no" may be ineffective as a teaching tool. Using our raised voice to influence behavior relies on a type of learning that is highly dependent on *timing.* The child needs to make an association between "no" and a specific misbehavior (stepping off the curb).

In contrast, if a child is older, and we catch him or her acting carelessly in the street, we can explain why we are upset, and our verbal explanations might influence future behavior. This type of verbal learning is controlled by different neural configurations, compared with the type of learning influenced by a loud "no."

The Contradictor

"You're contradicting yourself!" How many times have others used those words to describe our position during an argument? Although the brain is not irrevocably doomed to make silly errors of logic, it cannot avoid them altogether. Why? Because of what has been termed the "networking principle," especially the section concerning segregation.

If I said to you, "During the next ten minutes, list all the things you can think of that are green." *If* you agreed to do it, there is little doubt that you would come up with an impressive array of objects. However, if I provided more specifics, your list might be even larger: "Make a list of all green trees," "Make a list of all green insects," "Make a list of all green vegetables," "Make a list of all green fruits."

As the instructions become more specific, whole new categories of items are brought to mind, releasing a different flow of information. None of us can spontaneously access all the consciously available information within the brain. Yet, if the information is consciously available, why are we not able to just pour it all out? There are several reasons. First, the brain is constantly storing new material, so it would take years and years to clean out the closet, and even then, the job would never end because new material is constantly being added. Most important, though, information is stored contextually. If we are not reminded of a situation, we may not call forth spontaneously the details related to that situation. Finally, information is segregated and often stored in component parts. Strict rules must be followed to access even a fraction of what we know. At present, though, no one knows exactly how humans willfully "call up" information. Retrieval depends on processes that occur outside our awareness, although certain procedures are known to help. If we cannot remember someone's name right away, we can try to think about the person in a specific context, which might "jog" our memory. The brain clearly follows certain principles when it calls up stored information, but exactly how those principles work at the biological level are not known.

Each of us holds hundreds of beliefs and ideas within our heads, and many of those beliefs or ideas are diametrically opposed to one another, but they coexist peacefully, for the most part, because they are stored under different categories and are only called forth when needed (usually, one at a time). The brain's storage and information processing capabilities are highly situation-specific (i.e., segregated). Generally, we only recall that which is relevant for the task at hand and are spared from reviewing everything in our minds every time we need to remember something.

Thanks to the condition we experience as "consciousness," we can focus on only a limited number of ideas at once. Thus, it is perfectly natural for each of us to store opposing beliefs and attitudes yet rarely be confronted by those contradictions. I hasten to add, there are vast differences in this regard from one individual to the next. Some of us may be better at maintaining more or less consistent beliefs, others less so. One reason for individual differences may be that the brain has an almost unlimited ability to learn, so the amount of contradictory material we each hold depends, in part, on who we have learned from. Each of us has many teachers. "Teachers," in this sense, refers to all the influences that have contributed to the mammoth volumes of data stored in our brains.

With at least eighty billion information processing cells, and the volume of data that individuals accumulate over the course of a lifetime, the brain simply does not have the time, energy, or neural circuitry to compare every new piece of data with every other piece of stored information. Such a task would be impossible for even the most advanced computers, which can transmit information far more quickly than brains.

Because our brains segregate experience and information as a matter of course, the segregation of data can be counterbalanced, to some extent, through teaching and instruction. In the drug prevention field, to illustrate, knowledge of harmful effects of drugs, provided through education, may not carry over to a real-life situation in which a child is tempted to smoke. However, when children are given a chance to practice (through role-playing) how to say "no" in a variety of situations, they will have a better chance of refusing the temptations of drugs. In short, the downside of the brain's propensity to segregate information can be mitigated to some extent through training.

The Motivated Creature

We cannot remain unmotivated. To the contrary, our brains consist of several systems that support goal-directed behavior. Many of our "motivational" systems operate simultaneously and automatically and arise from different areas within our brains. Further, our brains have different ways of reinforcing us. We can experience pleasure from finishing a difficult work assignment or by eating when we are hungry.

Beyond the ability to be motivated by several "systems," there is also somewhat of a hierarchy among various motivational systems, and some psychological theories have sought to build on that hierarchy. One of the most famous American psychologists of the twentieth century, the late Abraham Maslow, wrote about different types of motivation. He separated "basic needs," such as hunger, from "metaneeds," such as justice.[32] Justice is uniquely human, whereas hunger as a motive is something we share with all animals. Each of these two types of "motivations" stems from different regions within the brain, and each may occur simultaneously or sequentially.

Different types or levels of motivation seem self-evident when we look at an example. A female college student, who is a single mother, puts the final touches on a term paper. During her work, the smoke detector lets out a shrill warning and she immediately smells smoke, then sees black smoke billowing from the kitchen. With the kitchen in flames, infused with fear, she grabs her child and gets out. Her fear originated from a lower and older part of her brain.

Once outside, as the fear begins to subside, she thinks about her term paper. Momentarily, she might even consider going back into the house, but decides against it. The thought of her term paper does not emanate from the brain region that produced the sensation of fear; rather, the thought of the paper, and the thought of returning, arose within the cerebral cortex, and as such, stemmed from a newer region of the brain.

Not only do motivations arise from separate regions of the brain, but as noted, motivation stems from many factors at once. I may desperately want an autograph of a basketball player and, therefore, choose to remain in line outside the stadium for several hours, in hopes of catching him or her after practice. The need to eat or sleep, though, would eventually intrude, to the point where I would have conflicting needs. People are constantly motivated by different regions within the brain, and the motivations may be compatible with one another, or they may compete with one another.

We should here revisit the notion of a network. In the foregoing example, it appeared as if different behaviors (the fear of fire) were *caused* by specific regions of the brain. Although emotions appear to emanate from the limbic system, the awareness of our feelings is registered by the neocortex. Complex events, like the apartment fire, are recorded and processed by many parts of the brain; and we have varying amounts of control over each of those processes. In other words, the degree to which we can directly influence various motivations depends to some extent on the motivation and the part of the brain in which it arises. Whereas I cannot instantly stop a feeling of fear, I can choose how I react to that feeling in many instances (though, not always).

The Brain Bypass

The more complex a behavior is, the greater the number of neurons that must be used to accomplish the task. We could define "complex behavior," in fact, as those behaviors or mental processes that require the greatest commitment of the brain's resources. The only reason this definition would be confusing is that we cannot see inside the head. We cannot be sure which activities require the greatest amount of resources. Generally, therefore, "complex" refers to performing many activities simultaneously, like answering the phone, turning down the volume on the stereo, checking the baby, and turning off the stove.

Now, reflect for a moment on your car and think of how often it needs servicing. Whereas a car's engine may have thousands of parts, our brains have billions, in the form of neurons, and those cells must work in concert with one another to accomplish a task. What is the likelihood, therefore, that some of them will break down, malfunction, or even die? Absolutely certain! Moreover, because our brains are biological, not mechanical, they are highly influenced by such things as food, oxygen, the time of day, fatigue, boredom, and so on.

In short, *all* of us experience some loss of brain functioning. Further, all of us are missing some capabilities because those capabilities never developed to begin with. This may seem disastrous, but it is not. It is normal. Most brain impairments are minor, so we may not recognize their presence. If we have never had perfect pitch, we are not likely to miss it. If we have always managed by counting on our fingers, we may never feel "mathematically challenged." We all have areas of strengths and weaknesses when it comes to brain functioning.

Because the brain is such a complex system, it is *sometimes* possible to *bypass* a specific area and end up with the same result that would have occurred in the absence of the deficit. If a young boy is having difficulty learning to write with a pencil and paper, his teacher might try using the blackboard, a typewriter, or computer keyboard. Each of these alternatives calls upon a different configuration of nerve pathways in the brain. The task of teaching the child critical thinking skills *may* still be accomplished by providing alternatives to paper and pencil. This strategy is one of the key approaches used in rehabilitation settings and in the field of special education. It is based on the proposition that any shift in movement, perception, or mental process causes a corresponding shift in the configuration of neurons that underlie the task. There are, however, many possible exceptions to this general strategy.

First, we can only "bypass" a deficit if it is limited to an avoidable area. A typewriter or computer keyboard is not going to help the boy if his problem stems from poor concentration or from an inability to tell one letter from another. The bypass strategy can work only if the deficit arises from a limited area that is not crucial for the principal lesson to be learned.

Second, as a general rule, younger brains are far more adaptive and "plastic," or moldable, than older ones. They can more readily adjust to deficits and call upon different configurations of nerve cells to compensate for a loss. When it comes to developing behavior compensations, youth has its advantages.

Third, if an entire brain area is damaged or underdeveloped, then the resulting loss is much greater, and compensatory strategies are less likely to be fully effective. If the damage or initial deficit is very small, then there is a much better chance of devising effective compensatory strategies.

Fourth, some capabilities cannot be compensated for if those capabilities did not develop in the first place. Someone who is tone deaf cannot acquire perfect pitch, even through hundreds of hours of piano practice.

Fifth, as a general rule, if a behavior is highly complex, like writing, there may be more choices on how to bypass a deficit, especially if the deficit is highly localized within the brain. In the writing example, each compensatory method allows the boy to learn to communicate with symbolic language, but each method calls upon different neural networks within the brain.

Finally, as mentioned, a simple-appearing behavior may, in fact, require a great deal of the brain's resources, while another behavior may seem complex but require only a small part of the brain. In one imaging study, people were asked to imagine themselves taking a walk through their neighborhoods, focusing (in imagination) on familiar aspects of their neighborhoods. The scientists who conducted the study reported widespread activation through various areas of the brain.[33] Generally, mental processes and behaviors require many types of skills, and only a careful examination by a trained professional can pinpoint (sometimes) where a particular deficit *might* arise within the brain.

CONCLUSION

The brain is a conglomeration of cells and structures, many of which originated during different periods of evolutionary history. *Each brain structure and each small cluster of cells provide a specific function that may not be exactly duplicated by any other group of cells.* For complex behaviors, such as talking on the phone and taking notes of the call, the brain may link "older" and "newer" areas together, through "executive" and "association" areas. In other words, the brain functions by forming temporary working alliances that often cut across regional and evolutionary lines. All of this is embodied in the major principle generated by this chapter, restated here:

PRINCIPLE 3: THE NETWORKING PRINCIPLE

Our brains are highly complex information-processing systems or networks that often segregate information. The concept of a "network" has two implications. First, for most behaviors and mental activities, such as thinking, many parts of our brains are temporarily linked together by brain areas called "association" and "executive" regions. Second, our brains are made up of many parts, some of which evolved during different periods of evolutionary history.

* * *

Whereas the brain is a complex network formed through millions of years of evolution, the brain has yet other attributes that provide key insights into our behavior. We will look at two of those next.

CHAPTER 4

THE HUNGRY AND BIASED BRAIN

From four to five billion years ago, our planet was born. Sometime thereafter, under the influence of the earth's environment and aeons of time, the first micro-organisms emerged. Beginning life as simple creatures, early life forms changed and evolved, and changed some more. Through all the changes, though, there was one constant: Once alive, all organisms retained remnants of their *biological* beginnings.

Because our brains are alive, they manifest a host of idiosyncratic ways of processing information—ways that mark our brains as both *biological* and *human*. Here we will focus on a few distinctive features of our brains that strongly influence how we experience the world, and afterward, we will draw some important inferences about behavior.

Often compared to computers, our brains and computers do share important properties. Information in a computer is segmented into different files. Within our brains, information is segregated or compartmentalized. Different types of memories, for example, are segregated from one another.

Computers rely on electrical energy to process information, and so do our brains. Neurons process information by relying on both electrical and chemical energy (we will look at these processes shortly).

I should not carry this computer analogy too far, however. To do so would lead to incorrect inferences about the brain and behavior. Our brains part company from computers in a number of fundamental ways that are highlighted in this chapter. This point bears remembering because our brains are often criticized when they fail to function as "efficiently" as computers, but the origins of brains and computers are quite different. The computer was designed by human engineers to solve specific types of problems. Various brain functions are believed to have been naturally selected for specific environments.

To begin our look at the brain's information-processing capabilities, we must look at the most basic unit of processing, a single cell. From an evolutionary standpoint, prototypes of the cells within the brain predate any single brain structure by millions of years.

BRAIN CELLS

The number of neurons in our brains is extraordinarily large, estimated to be eighty billion or more.[1] However, the exact number is far less important than the general idea that estimated numbers are too huge for many of us to fully grasp, a situation that is analogous to comprehending the number of stars in the sky or the number of grains of sand on a beach.

The eighty billion estimated cells only pertain to those that process data. At least half of the brain's cellular mass is comprised of cells that perform no information-processing functions. These support cells are called *glial cells*, of which there are different types. The various types, collectively, provide such crucial functions as holding nerve cells in place, cleaning up debris and waste matter, and insulating *some* neurons, much like the coating of an extension cord. These are only *some* of the functions of glial cells, but it should be emphasized that they are essential for life itself.[2] However, since they perform no information processing, per se, we should refocus our attention on the real stars of the show: neurons.

All the neurons we will ever have are generated during a period called "neurogenesis," which occurs *before birth*. At times during neurogenesis, fifty thousand nerve cells are generated each second! During this phase of brain development, cells "migrate" to their respective places within our nervous system. A single layer of cells eventually becomes the multi-layered brain.[3] Within approximately eighteen weeks of prenatal development, all neurons of the human cerebral cortex have developed and traveled to their respective locations.[4]

At some point during early brain development, the number of neurons within the brain is reduced from an extraordinarily large number to *a mere eighty billion*! Experts now believe that an *optimal* number of cells is necessary for peak efficiency (not just the largest possible number), but the discovery of this elimination process is too recent for experts to know exactly why the brain eliminates some cells after they have been formed.[5]

Although our brains have billions of neurons, once a neuron dies, it will not be replaced. There may be some minor exceptions to this general rule, but overall, it is probably best to conclude that we are born with all the neurons we are ever going to have. Scientists are experimenting with ways of regenerating lost neural tissue, but for all practical purposes, when neurons are destroyed, they are gone forever.

Perhaps now is a good time to take a minor detour to discuss a common myth about the brain. For most of my life, I have heard comments to the effect that people use only about 10 percent of their brains (or some such figure). Without doubt, our brains can compensate for some injuries or deficits, especially among the very young. Furthermore, all of us can learn throughout our lives. These facts would suggest that we have some reserve or backup capacities. However, it is *very unlikely* that the brain has anything close to a huge capacity that is not being utilized.

Our skulls are very crowded. Because the brain evolved in response to environmental needs, it is unlikely that extra cells evolved that were not needed for some function. Thus, it is probably safest to conclude that most of us could maximize our potential to some extent, but it seems unlikely that we have large numbers of brain cells that are not serving a function.

While on the general topic of "unused" brain cells, I will digress a bit further and mention a related topic that is less speculative. The consistency of our brains is similar to that of jelly. Our brains are very fragile. In spite of this fact, we can easily find examples of individuals who subject their heads (and brains) to enormous risks.

Some years back, I attended a conference on closed head injuries (injuries to the brain in which the skull was not fractured), and one of the presenters was a neurologist who estimated that even *minor* blows to the head probably result in some loss of functioning. I raise this point here because of the often expressed idea that we have cells to spare. In fact, the opposite *may* be the case, based on the number of people in this country alone who have neurological impairments. While compiling research studies for this chapter, I learned that roughly fifty million people in the United States lead lives that are "diminished" by damage to the brain or spinal cord and an estimated one million people experience head or spinal cord injuries each year.[6] Facts like these serve to illustrate how vulnerable our brains really are.

There are many types of neurons. The location within the brain (and the cell's genetic code) determines their function.[7] Cells are specialized and adapted for different jobs. Some convey colors or sounds. Some convey information about movement, and others perform integrative or executive functions. When information about color first enters the brain, that information is processed by a specialized type of cell. In all likelihood, a cell that registers color does not react to sounds or other types of information. Individual properties, like color, are then integrated with other properties, like size, by groups of cells that perform integrative functions. Once information has been integrated, other cells attribute meaning to the developing picture. The attribution of meaning is accomplished by cells within "executive" regions of the brain (the term executive is figurative, denoting those brain functions that involve complex decision making). At present, it is not known how the brain transforms the actions of millions of neurons and derives meaning from that activity.

Regardless of how different cells look or what specific functions they serve, all neurons have certain properties in common, and a general overview of those properties is germane to this chapter.

A neuron consists of a cell body, dendrites, and an axon.[8] In turn, each of these parts consists of still further divisions, but only scant details will be discussed here in order to maintain clarity. In addition to the "parts" of a single cell, the connecting point between two cells is called a synapse, and it, too, will be discussed here.

The Cell Body

The cell body or *soma* houses the cell's genetic makeup, which, in turn, determines the cell's function within the nervous system. The soma also contains the biological apparatus needed to keep the cell alive, including the ability to provide energy, manufacture various chemicals, and eliminate waste products.[9] *Very importantly*, all cells require energy to function and remain alive, and energy comes primarily from glucose.

Dendrites

The word *dendrite* comes from the Greek word for tree, and dendrites often appear like branching tree limbs, hence the name.[10] Dendrites reach out well beyond the cell body and *receive incoming* messages from other cells, usually cells that reside in close proximity (but not invariably). Because of its branching dendrites, a single nerve cell may receive thousands of messages from other cells. Furthermore, dendrites can grow and change throughout our lives, so although brain cells cannot regenerate, the dendrites of a cell can increase or decrease. This capability is thought to contribute to our ability to learn.[11]

Axon

An *axon* is the part of a cell that conducts impulses away from the cell body. That is, one end of the cell body is elongated and stretches out like a long tree root. At the end, the axon may branch into small twiglike structures. Axons are often covered with fatty glial cells that insulate the axon to speed up the transmission of outgoing signals. This covering is called *myelin*.[12]

Synapse

Synapses are minute spaces between two nerve cells. The term synapse generally refers to the microscopic space between two cells, but includes the specialized membranes of both sending and receiving cells.[13]

CELLULAR MESSAGES

When messages are sent through the brain, sending cells emit a chemical into the synapse called a *neurotransmitter*, and receiving cells "pick up" or receive the neurotransmitter.[14] Figuratively speaking, incoming "messages" to a cell determine whether or not it will "fire." When cells fire, they are sufficiently stimulated to send a signal to another cell. Specifically, if the combined strength of incoming signals is strong enough, a signal will be sent down the line. This is the simple version, but some additional clarification is needed.

Cell bodies and their axons transmit information through the use of electricity that is generated by positive and negative ions. Although this part of the signal is electrical, it is by no means instant, compared to the flow of electricity through an extension cord. The fastest nerve cells may transmit information at approximately three hundred feet (100 m) per second, but compared to the flow of electricity in a wire, this is a snail's pace.[15]

When electrical impulses arrive at the end of the axon, the continuation of the signal is no longer electrical, it becomes chemical. Neurotransmitters are released into the synapse and "picked up" by the membrane of a receiving cell.

Recently, scientists have discovered that neurotransmitters can influence cell genes that produce long-term alterations in synaptic activity. This is extremely important because our brains not only transmit information, analogous to a telephone line, but they also change in response to use. Stated another way, our brains adapt and change, and it is likely that learning is the result of synaptic changes or interactions with neurotransmitters.[16] Only a handful of neurotransmitters had been discovered until recently, but now over one hundred have been identified, even though scientists do not fully understand the function of each.

Very broadly, beyond changing a synaptic connection, neurotransmitters either "excite" a cell, causing a signal to be sent, or "inhibit" a cell, preventing a signal from being sent.

Since a single cell might receive thousands of incoming messages, whether the cell fires or not depends on the added effects of all incoming signals. If fifty inhibitory messages and fifty excitatory messages are received at the same time, the receiving cell will not fire. The transmission of a message is the summation of incoming signals.

At the cellular level, a message does not contain information, per se. An analogy might be to the single note of a single instrument within an orchestra. A musical piece only becomes recognizable when all the notes are combined over time.

How does the brain combine millions of "go" or "no-go" messages, and from that, extract meaning and make judgments? This is the billion dollar question. Clearly neurotransmitters are central to the flow of information in the brain, as are synaptic connections, but beyond this general knowledge, no one knows how the brain derives meaning and makes decisions, based on the actions of millions of neurons working in concert with one another. At present, scientists can provide only theoretical models for how the brain integrates information and make *rough* estimates where those functions *might* occur.

Since even the simplest processing task may require hundreds of thousands of cells, even the fastest processing takes *time*. Relatively speaking, therefore, our coordinated movements are slow. Electrochemical signals take measurable time, so when we consider that enormous numbers of cells are involved in information processing, we can appreciate why the brain is slower than a computer.

Even when brain cells are not processing information, they still maintain minimal metabolic processes to remain alive. However, when cells become actively involved in data processing, their metabolism increases (they require more fuel). This property of neurons has helped scientists understand them. Since working neurons require more fuel than resting ones, brain imaging techniques were developed to capitalize on this fact. Positron emission tomography (PET) is one such technique. The person to be analyzed swallows a glucose solution that contains a short-acting radio-isotope. (Because the radioisotope is short-lived, it poses no harm to the person swallowing it.) After the isotope has been ingested, the person is asked to engage in a variety of mental tasks, such as visualizing a certain word or speaking the word out loud. There will be an increase in activity in those cells required for the task. A picture of the blood flow can then be generated because the isotopes make the active cells stand out from cells that are not working.[17] (The very fact that PET works is proof that active neurons require more energy.)

PET scans are fairly slow and require sophisticated equipment. Tasks that can be used in PET studies require people to repeat simple procedures over and over, like repeating a single word or solving a simple mental calculation. Therefore, PET cannot measure highly complex brain pro-cesses or actions that we perform in our natural environments. In short, even the most sophisticated imaging technology, thus far, can focus only on the simplest of mental processes. Interestingly, PET studies have shown that simple problem solving requires more energy than passive viewing.[18] (We probably knew this already, but it takes more energy to solve a problem than watch television.)

When the foregoing discussion about the energy requirements of brain cells is combined with the networking principle, an interesting possi-bility arises. In describing information-processing systems, Daniel Gilbert of the University of Texas at Austin, argues that "no mental system, whether natural or artificial, has unlimited processing resources."[19] For cells to work, they require both energy and time. However, because *no* mental system has unlimited resources, the obvious conclusion is that our brains are influenced by this fact. Gilbert suggests that our brains must make choices when solving problems. If there is not enough fuel to do everything, something must be sacrificed. (This conclusion has logical appeal, but it is not without its critics.)

The brain's need for time and energy to process information takes on greater significance if we consider the prehistoric environment that shaped brain development. In order for the brain to ensure the survival of individ-uals within a dangerous world, the brain has to process information rather quickly. However, as we have seen, built-in biological constraints set upper limits on how quickly data can be processed. We will put this in perspective when we reconsider what prehistoric life must have been like when the first modern brain appeared, perhaps two hundred thousand years ago.

At that time, the world had many large predators, and all humans were hunters and gatherers. None of our prehistoric ancestors, however, had the advantages of modern weapons or protective clothing. No automatic weapons or assault rifles for them, not even the use of primitive metal blades, and many may not have had the advantage of fire. With little doubt, prehistoric humans lived in a world filled with danger, and they could have been prey just as easily as being predator.

On the one hand, as we have seen, the brain's processing took both time and energy. Survival, though, would have favored those who could react quickly. Individuals who demanded absolute accuracy, or those who were too slow, probably did not survive. The brains that evolved had to juggle the demands of being "accurate enough" with the limits placed on processing by both time and fuel constraints. Therefore, in all likelihood, natural selection favored a brain that could derive a quick, *adequate* solution, in contrast to a brain that was unfailingly accurate. In the face of a charging animal, the exact nature of the beast was far less important than a quick response.

The need to balance fuel and to devise a quick but adequate solution provides a useful way of considering the brain. To balance competing needs, the brain takes all sorts of shortcuts when it processes information, and absolute accuracy is not one of its strengths. An example may clarify this. If I had experienced only unpleasant encounters with large dogs in the past, the next time I saw a large dog, I would react quickly with some degree of apprehension. Having a quick response at hand, *based on past experience*, can be very useful. If the new dog turned out to be vicious, my quick conclusion about its nature could save my life. If my conclusion was mistaken, the cost to me would be minimal. Our brains place a premium on "previous solutions" and recycle those whenever possible, *often without awareness of when or how those solutions came into existence.*

Much, if not most, of the brain's functions occur outside of our awareness. We have probably all heard about "the unconscious." The unconscious is a theoretical idea, often associated with Sigmund Freud, who, in his earliest writings, emphasized the importance of the unconscious, but later refined the concept considerably.[20] In light of recent brain research and modern imaging studies, it seems clear that *most* brain functions unfold outside of our awareness, but today's scientists have a better understanding of why.[21] The concept of the unconscious, as Freud described it, has been criticized by many researchers, but the types of behaviors that Freud observed are nevertheless quite valid. People often behave on the basis of influences that occur outside of their awareness.

The term "unconscious" is widely used in our culture, and most of us think of it as those thinking or mental processes that occur outside of our awareness. Nevertheless, we will use the term nonconscious in most instances, because it does not carry Freudian connotations, and to some professionals, that distinction is very important. Both terms, however, refer to those processes that unfold outside of our awareness.

It is now known that many learning processes may begin in our awareness, but are moved to other parts of the brain, possibly to save energy or free-up our attention. For example, some types of learning are switched to different parts of the brain after a challenge has been mastered. When we learn to swim, we may consciously focus on each aspect of a swimming stoke, but when swimming is mastered, it unfolds effortlessly, often without our conscious effort "to swim." One of the possible reasons for this switch is to save energy. "[T]he shift of learned material to a different part of the brain is like putting the information on 'autopilot' and it appears to be the brain's way of conserving energy to learn new things."[22]

Traditionally, human thinking has been described with such words as "rational," with the assumption that our thinking should follow formal rules of logic. However, most thinking occurs nonconsciously; it is automatic; it is rarely open to direct study; and it does not follow recognized rules of logic. Our thinking is context-specific and very practical. The primary operating directive for the brain is to find a quick serviceable solution and move on. The less thought required, the better, from the standpoint of time and fuel consumption.[23]

Since the brain has survived to the present, we can assume that is has achieved a good balance between fuel, time constraints, and passable accuracy. However, this balancing act unquestionably influences the ways in which we process information. Here are some of the idiosyncratic ways our brains make sense of the world.

Generally, our brains show a marked bias toward believing (as opposed to disbelieving). "Acceptance and rejection are not merely alternative outcomes of a single assessment process, but rather, acceptance is psychologically prior."[24] It is much quicker for the brain to have a built-in bias when evaluating new information than be forced to fully consider all the possibilities. This is a broad statement, so before we pass judgment, we should consider a few more details.

Certainly, if we were to go out and conduct a random poll with a questionnaire containing only "yes" or "no" answers, we would be unlikely to find any evidence of this thinking bias. Thus, the brain's "tendency to believe" needs to be placed in perspective.

First, children have a clear tendency to believe what they are told, especially when they are told by someone they trust; similarly, they have a bias toward believing what they see.[25] To a three-year-old, if a pencil immersed in a glass of water looks bent, the pencil is bent, even if a straight pencil is put into the water in the child's full view. Below the age of four or five, children will believe what they see and are told, and we can speculate why.

If prehistoric parents were to have yelled at their children, "Watch out!," children who did not "believe" would have perished. Belief and disbelief are not on an equal par, so natural selection may have favored belief over disbelief as a first response. (Also, belief may take less fuel than disbelief. This is because we may need to understand a proposition first before we reject it, so rejecting it is a second phase of thinking.)

Similarly, younger children are less sophisticated than older ones. Their thinking is more "concrete," which means they are unable to consider abstract ideas until they are older. Even children who live in houses with no chimneys will believe that Santa can enter their homes through some magical means if they are told that Santa exists. Accepting a literal statement at face value is readily accomplished by young brains; it is much easier than not taking the statement at face value.

Furthermore, not all messages are the same. For some types of message, we have no choice but to believe. To a child who has been raised with television, seeing a picture of a new four-wheel drive vehicle in the wilderness is a type of message (there are four-wheel drive vehicles in the wilderness). The young brain has no means of analyzing the imagery and knowing that cars and wildernesses do not always go together. The imagery is the message, and seeing is automatically believing.

Even among adults, there are certain types of messages that we might assimilate and "believe" almost automatically. We can watch someone solve a simple problem. Afterward, we can effortlessly copy their behavior. Copying someone else's behavior, though, is a form of believing; it is a form of assimilating information without question.

As mature adults, our brains appear to have a processing bias for simple, concrete messages, as opposed to messages that are more complex. Simple messages take less fuel and less time to process. They do not require high levels of motivation to understand. Thus, simple messages have a better chance of being *understood*. Being understood, in turn, provides the message with an edge over messages that are not understood. There is little doubt that a simple barb, like a one-liner during a political campaign, *is* easier to understand in contrast to a lengthy explanation of a technical nature; hence, the sound bite. (Of course, this assumes that the simple message is at least plausible because a ludicrous message takes little energy to see through.) All of us can quickly assimilate the message, "He's a crook," but it takes a great deal of time and energy to analyze complex messages, and perhaps consider another explanation for a behavior that *appears* to be dishonest, but may not be. Averaged over large groups of people, simple is easier, if for no other reason than people have to be highly motivated to wade through highly complex messages. (Clearly, there are exceptions to this generalization, but averaged over large groups of people, simple does appear to have an advantage over the complex, relative to believability.)

As we learned earlier, most of the brain's processing occurs outside our awareness, nonconsciously. Nonconscious processes, in turn, are heavily biased toward sensory impressions, like a strong feeling about something without awareness of where the feeling came from. Nonconscious processes, too, are often concrete and literal, partly because our nonconscious attitudes are formed early in life, during a time of relative immaturity, when the child's brain interprets events literally and concretely.

Let me give a good example of a concrete message taken from clinical experience. We might ask someone, "What brought you to the clinic today?" Someone who interprets the message "concretely" would say, "a car." Most of us, though, would see that the question is meant figuratively: "What problem brought you in today?" (We'll revisit the idea of nonconscious attitudes in chapter 10.)

The likely bias of concrete or simple messages over the abstract has many ramifications, but there is a difference between the two types of messages. "that which is not" is an abstraction and "that which is" is concrete and more tangible. The latter is easier for us to comprehend and, therefore, has a greater chance of being understood and often has a better chance of being believed. When these two types of messages are pitted against each other, the concrete has a decided advantage. When a bias toward the concrete is multiplied by large groups of people, then we begin to see social effects. Short-term, tangible solutions to social problems have a higher chance of winning over more complex, long-term solutions, which are more abstract because they are in the future. Our needs are more real (tangible) than the needs of "future generations" (which is an abstraction).

The foregoing discussion has established the fourth of the twelve general principles.

PRINCIPLE 4: THE AUTOMATIC PRINCIPLE

In comparison to a computer, our brains process information rather slowly. However, since our brains evolved in a very dangerous world, there was a need for quick action. Thus, to compensate for slow speed, our brains rely on automatic reactions as much as possible because automatic reactions are quicker. Automatic processes, though, often neglect important information in the name of expediency, with the result that our "thinking" is always biased and sometimes wrong.

This principle should *not* be interpreted as a criticism. The brain was never designed to function like a computer, even though it is often compared to one. The brain's biases are thought to be naturally selected because they provided an adaptive edge. It takes too much energy and time for the brain to evaluate every situation anew, so experience from the past allows us to make quick, automatic judgments.

Aside from the need to juggle the constraints imposed by energy and time, along with the need for adequate accuracy, there is another property of the brain that is uniquely human and also has many implications for understanding behavior.

THE MEANING OF MEANING

Our eyes do not perceive all wave lengths of light, and our ears cannot hear every pitch. Our skin cannot perceive all temperature changes (nor even tolerate others). Clearly, there are limits beyond which our senses cannot go. Compared to some animals, our eyes seem weak. Similarly, the rich sounds and smells that inundate Fido's world lie beyond our grasp. Nevertheless, when our brains integrate various sensory inputs, we *experience* a picture of the world that seems unerringly accurate.

The limitations of our perceptual and thinking capabilities have been known for a long time. Perhaps one of the driving forces behind the invention of science was the recognition that our most fervent beliefs could still be wrong. After all, for thousands of years people assumed the earth was flat, and they were wrong.

Rather than being a handicap, the limits of our sensory organs provide us with tremendous advantages that were probably naturally selected. There is too much going on for us to pay attention to it all, much less process and make meaning of it. If our brains *had* to rivet their attention on every single insect, every movement, every light, every sound, and every change in temperature, we would be effectively paralyzed.

Fortunately, our brains attend to only a fraction of the events going on around us, usually those things that are *personally relevant* at any given moment. If we are hungry, a restaurant we have never seen before might "jump out" at us, but if we are excited about a new job promotion, the restaurant may escape our notice. Our focus is ever-shifting, depending on our needs or goals at the moment. As wave lengths of light or sounds are sent to the brain and integrated into objects or events, only some of those integrated stimuli will be tagged with special meaning, only those that relate to our current needs.

If I am an entomologist and see an extremely rare butterfly, my attention will be riveted to it until I can devise a way to study it more closely. On the other hand, if I do not know one butterfly from the next, I may not notice the butterfly. This example illustrates that personal relevance is often based on the meaning we give to events. However, the meaning we give to events influences our perception of those events. A rose ceases to be mere wave lengths of light and a special shape with a distinctive odor when it is labeled a "rose." At the level of conscious recognition, our brains have already integrated incoming and diverse information and tagged that information with a label: "This is a rose." In assigning labels, however, we have gone well beyond what our senses first reported to our brains.

In addition to selecting what to attend to and attaching meaning to it, our brains have a built-in, overriding bias for events that appear threatening or potentially threatening. Both negative or dangerous-appearing events take precedence over neutral or positive events. This phenomenon has been called "automatic vigilance."[26] If we are standing in line at a department store and get a whiff of smoke, everything is forgotten except the smoke, until we determine that the smell is coming from the building's ventilation system and realize the smell is already dissipating.

We do not recognize negative events faster than anything else (recall that the brain's processing speed is relatively fixed), rather, when many things are going on at once, information about negative or potentially harmful situations is more compelling to our brains. All other factors being equal, negative stimuli have more power to capture our attention than positive or neutral stimuli.

We can now consider the next general principle.

PRINCIPLE 5: THE HUMAN FACTOR PRINCIPLE

One of the principal characteristics of being human is that we use symbols to interpret the world. However, by reducing events to symbols, be they words or numbers, we alter our perceptions and memories of the very events we are thinking about or remembering.

Principle 5 pertains to many phenomena, like memory, learning, and thinking. Because our world is too complex for us to record all that is going on, this complexity, coupled with the biological constraints we noted earlier, force our brains to judiciously pick and choose what will be attended to and what will be ignored.

The tendency of our brains to pick and choose, though, is certainly not random. Our brains demonstrate some predictable biases in how they process information—a theme to which we will return shortly.

IMPLICATIONS FOR UNDERSTANDING BEHAVIOR

Personal Needs Influence Meaning

We *behave* on the basis of how we interpret a situation. As we have seen, though, our interpretations are often influenced by meaning, and it is we, the perceivers, who attribute meaning. To illustrate, studies of aggressive boys have shown that they are more likely to see others as being aggressive, and therefore an ambiguous event, like an accidental bump in the hall, is more likely to evoke aggression among boys who are already aggressive.[27] In other words, the boy had to interpret the event before responding.

Consider another example to help clarify this. Two different people are sitting at a picnic table. One is a trained biologist who specializes in the study of spiders, the other is a psychologist who knows little about spiders. As they sit and eat their lunches, a spider crawls onto the table. The biologist notices the spider and immediately knows that it is a common, harmless variety. The psychologist, who knows nothing about spiders, becomes frightened and completely loses track of the conversation. The biologist notices the psychologist's anxiety and begins to laugh, as the psychologist starts looking for something with which to kill the spider. This scene illustrates a crucial point: We, as humans, attribute meaning to events; the meaning is not inherent in the event. We then act on the meaning we have given to the event. This scene also illustrates that the meaning we give to an event can be at odds with "scientific reality" or with the interpretations of others.

The Stereotype

Stereotypes are now considered by scientists who study thinking to be one of our brains' many ways of saving energy and quickening response time.[28] As humans, we may have little choice but to stereotype. To consciously consider most objects or events, often we must first attach a symbol or label. This is how we give meaning to people and events. However, it is obvious that we can only apply labels or symbols with which we are already familiar. Despite the damaging consequences of some negative stereotypes, research has nevertheless demonstrated that stereotyping is very widespread. In fact, our stereotypes are often highly resistant to change, *even* when we are provided with a great deal of contradictory information relative to the beliefs we already hold.[29]

One likely cause of negative or harmful stereotyping is that labels are applied to someone and then negative inferences are made on the basis of those labels. Although we can acknowledge our own biases, to some extent, to do so is not the first response of our brains. The first response is governed by the needs to save fuel and time and to devise an adequate solution quickly, usually by relying on an "old" solution.

If we consider the brain in prehistory, humans lived in small hunting and gathering bands that probably had little contact with other bands. The brain's tendency to stereotype may have had few negative repercussions under those circumstances. In fact, stereotyping may have had important survival advantages over and above the virtue of saving fuel. It may have been in a band's best interest for its members to see the world, and perhaps strangers, in a stereotypical manner because similar views could have helped provide solidarity within the group. The brain's bias toward stereotyping, which is probably hundreds of thousands of years old, may have far different consequences in today's world than it did during human prehistory.

In sum, our brains constantly scan the world. Mostly, we look for information that confirms what we already believe. Then, usually on the basis of past experience, we look for a match between the past and what is going on now. We make comparisons quickly based on the most obvious characteristic of a new situation, typically drawing upon previously used responses or conclusions. This is the brain's *first* response system, and even trained scientists may not apply their training outside specific areas of expertise. For any of us to analyze our thinking or beliefs, we have to be motivated to do so because it takes more energy to ponder our own conclusions. However, if we *think* we have solved the problem at hand, then we are not motivated to look farther for a solution.

When Old Solutions Fail

Since the brain's first line of attack is to look for an old solution and apply it to a new situation, we might wonder what happens when our brains encounter situations that have never been encountered before. This happens all the time.

The first thing to occur is that processing is slowed down, sometimes way down. One characteristic of an emergency, in fact, is that it is so far out of our normal experience that we are overwhelmed by the demands of the situation. (If we are highly trained for such an emergency, the challenges are less formidable.) The main point is that new and highly unusual situations often leave us without any ability to cope, at least temporarily.

I had a very interesting experience some years back at a place where I worked, which illustrates the foregoing point very well (although I am somewhat embarrassed to admit what happened). I arrived for work one day at a mental health center. There was a car parked outside the front door, with its engine running and the driver's door wide open. I did not approach the car directly, rather, I walked around it slowly as I approached the front door of the building. When I arrived at the door, I noticed that it was open and there was broken glass everywhere. Then I noticed that the front window of the building had been broken. Although you can probably figure out what was happening, I was so used to going to work and opening up the main door, I could not separate myself from old ways of viewing the situation. Since I was a therapist, the only thing I could piece together was, "Someone must need help." (That "someone" was me!)

As it turned out, the building was in the process of being burglarized. That thought never occurred to me, initially, not until I was met in the front lobby by a man with a tire iron, and even then, I was confused. Since I had never previously been subjected to theft, assault, or burglary, I could not draw on past experience to put the facts together. (Fortunately for me, the man ran past me, jumped into his car, and sped off.)

Since our brains look for old solutions to solve new problems, only when old solutions fail do we start to consider alternatives. From an adaptive perspective, this makes sense. A completely new solution takes a great deal of time to generate. In the break-in situation, my reaction was very slow because I had to integrate new information completely outside my previous experience (and I should add, in highly unusual situations, we are likely to be more emotional, and high levels of emotion *can* also impede our thought processes).

Overuse Is Believing

Just as old solutions most often form the basis of our thoughts, the frequency of our past thoughts determine how readily we will call upon them again. If we call on some ideas or formulas all the time, we are more likely to use those ideas for future reference. Thus, the relevance of new information may be less important than previously used information.

When a racist has practiced hating certain minority groups for a lifetime, a racist epithet comes easily to mind and is used to explain the behavior of a hated racial minority. Even if the minority member is behaving like everyone else, a reliance on past stereotypes may blind the racist to what is actually going on. This is one reason why negative, racial stereotypes are so difficult to combat. The brain is far more likely to impose old meanings on a new situation than rethink situations. People have to be both taught and motivated to look for errors in their conclusions, and unfortunately, not all people are. (Moreover, as noted before, when problems appear to have been solved, we lose motivation to look for other solutions.)

Uncommon Creativity

How many creative ideas do we have each day? I am not aware of pertinent data on the topic because of the difficulty of defining the term "creativity" and marking its occurrence. However, from personal experience, I know that I am far more likely to draw on that which "I already know" or think I know, than revise my thinking. It is often very difficult for us to perceive old situations in a different light, unless someone points out an aspect that we have not seen before.

Creativity, though, often involves the ability to look at old situations and put the pieces together in ways that are completely novel. Likewise, scientific breakthroughs often come from the ability to take old puzzles and put the pieces together in ways that have not been tried before. One reason that creativity may be relatively rare is that creative processes require us to go beyond the brain's normal ways of processing information. I should add, though, once people get into the habit of being creative, like any other habit, creativity may become more likely, it may come more easily.

The Attention-Getting Power of Danger

As we learned, our brains are exquisitely attuned to the negative, which includes outright threat, dangerous situations, and probably negative innuendos or implied threat. One of my favorite examples is that of insurance company ads on television. From a theoretical, brain-based standpoint, those ads are brilliantly conceived. The general scenario usually unfolds something like this: We are shown a warm, loving family, either in the safety of their home or car during the opening scene, followed by a dramatic portrayal of some disaster, a fire or car accident, followed by the assertion that you and your family are cushioned from disaster if you have insurance. Now, admittedly, I have probably missed some of the finer points of these commercials, but the overall influence is based on the brain's attention to danger. The brain's bias toward danger is well documented, and that bias can be easily capitalized on to motivate people (in this case, to buy insurance).

Simple Is Easier

Simple ideas have a greater chance of being believed than complex ones, generally. It is easier to say that a plane flies because the wings provide lift, than explain the principles of aerodynamics. This "simple is easier" principle has all sorts of social implications. For example, in a political campaign, if someone can come up with a plausible, but simple explanation that *appears* reasonable, in comparison to a complex, technical solution, the "simple" has a built-in advantage. If a politician in the same campaign can come up with a simple adage that is also negative, and use that to describe an opponent, then so much the better. Combining simple messages with negative ones are often very effective, and these messages are difficult to counter.

Simple takes priority over complex over time. If a simple idea explains a phenomenon, or appears to, then someone will not be motivated to look deeper into a problem. Thus, if a student is considered to be "lazy," then other explanations for school failure may not be sought. Conversely, the more complex the real reason might be, the less likely that reason is to be discovered, because it will take a great deal of time and energy on someone's part to uncover the complex reason.

This tendency to stop at a simple answer, however, can be mitigated in two ways. First, we can accept the possibility that our hypotheses might be incorrect, which provides additional motivation to look for alternative explanations. Second, we can actually check our assumptions by testing them. As we explored earlier, each of us employs scientific processes from time to time. We can devise simple experiments to see if our conclusions are valid. In fact, one of the simplest "experiments" is to seek the opinions of others and see if they agree or disagree with conclusions we have already reached. By doing so, we may discover an aspect of some problem that we had not seen or anticipated.

Believing Is Acting

We experience our picture of the world, by and large, as accurate, most likely because there was an adaptive advantage for us to believe what our senses told us. Because the picture we derive of the world seems real to us, it follows that we will act on our perceptions and conclusions.

So, if I am hungry, and I am driving around in search of a specific type of restaurant, I have essentially programmed myself to find a restaurant of a certain type. With my focus on restaurant signs, the last thing in my mind is someone on a bicycle. I might drive within a few inches of a cyclist and not even see the person.

We have probably heard accounts of accidents where the driver said, "I never saw him." That explanation may not be as improbable as it sounds. Even though our eyes may have registered the light waves given off by a cyclist, the brain may not have been "programmed" to give meaning to a cyclist, since the need was to find a restaurant.

Meaning Precedes Memory

We do not remember all raw sensory data, if any, partly because there is too much going on at any given time for us to store it all. Thus, we remember events only after our brains have assigned meaning to them. What happens, then, if our brains assign an incorrect meaning? I once had a client tell me that he was at a cemetery as a boy of three, for a funeral. He still maintains a vivid recollection of seeing a ghost. However, three-year-olds are capable of seeing and believing things that you or I would interpret differently. A three-year-old, for example, can believe in Santa Claus, but when we see "Santa" as adults, we recognize this as an adult in a costume. When children form memories early in life, they store those memories on the basis of their age, and some of those memories persist for life, and the original meaning assigned by a three-year-old brain may not have been updated. (What makes this point more significant is that past memories influence how we interpret new situations.)

Each of us, in effect, may grow more certain of our worldviews, partly because, as we form memories, we subsequently interpret new events through old memories. For some of us, the world is a wonderful, loving, safe place; for others it is a dangerous and hostile world. As we each filter the landscape through preexisting beliefs and memories, we are more likely to see things that conform to preexisting worldviews and less likely to see things anew. Memories not only store experience after we have interpreted that experience, but memories also provide us with a pair of colored lenses through which we see the world.

"Not Now, I Am Too Busy"

To draw upon a computer analogy, the brain does not have unlimited "multitasking" capabilities. Too many demands cause our brain to juggle output. Since our brains control both vital functions and behaviors that are comparatively discretionary, there are certain activities that absolutely *cannot* be short-changed, like the regulation of heart beat, body temperature, and so on. Thus, if you are running a little low on food or oxygen, possibly *because* you are running, some information-processing or less vital regulatory functions will be shortened or eliminated. Recently, I rediscovered this simple principle by accident.

At the local gym, I was engaging in a little mind game to make the treadmill less boring. I was attempting to compute how much torture I still had to endure by subtracting the number of minutes completed from an overall goal. As my heart rate went higher and higher, the ability to do even simple subtraction became almost impossible. Eventually, I had to use my fingers for counting, and I still made mistakes.

In a mental health clinic at which I worked, I was seeing a male client whose wife insisted that he seek treatment for his sexual obsessions. She told me that he talked about nothing but sex, that he stashed "dirty" magazines all over the house, and that he frequently propositioned female neighbors. The degree of his obsession about sexual matters, if true, definitely seemed abnormal.

This man also had chronic heart problems, and while he was in therapy, his doctors learned that the blood supply to his brain was severely compromised. Emergency surgery helped to correct the problem, and almost immediately afterward, his sexually inappropriate talk and behavior significantly diminished. He was transformed into a different person—all because the oxygen to his brain was restored to a normal level.

"You Are in Denial"

The concept of "denial" is widespread in our culture. Generally, people use the term to imply that someone is blocking essential information from his or her mind, or ignoring that information. There is often the insinuation of purpose, like, "He is in denial about his health," meaning he has a potential health problem that he will not address.

As originally formulated, denial was thought to protect individuals from anxiety-producing realizations. Denial was also thought to require an expenditure of energy to keep objectional material outside of awareness (in the "unconscious").

However, based on the principles presented in this chapter, an alternative explanation for "denial" is not only plausible, but in my view, more likely. Each of us selectively attends to any situation in light of our personal needs, so what appears like denial to others may only be the way our brains select and choose what is personally relevant at any given moment. Consider what our prehistoric ancestors had to do to survive; they had to maintain a high degree of awareness about what was going on around them *in the moment*—whether they were hunting or avoiding dangerous animals. It seems unlikely that they would have survived if they had brain mechanisms that enabled them to "ignore" important information as a means of warding off anxiety. Rather, a more likely explanation for what we call "denial" is that, at any given time, we are focused on what is immediately going on around us, *often* to the exclusion of distant possibilities. Thus, in the case of someone who smokes and is in denial about the health risks, it is more likely that the person's immediate focus is on matters other than long-term health risks. (Earlier, we discussed the concept of "automatic vigilance," in which the brain pays more attention to the negative. However, when possibilities are too distant in the future, they do not appear real, so to the brain, there is no threat.)

There is no doubt that people engage in thinking mechanisms that help them ward off fear or anxiety. We often divert our attention from something that is unpleasant. However, the tendency to attend to some aspects of a situation and ignore others is a well-documented human characteristic that should not always be considered to have a defensive purpose. Our brains are masters at shifting focus, depending on our current needs. We are highly dependent on moment-to-moment changes within the environment, and the brain places more emphasis on immediate outcome than future possibilities.

CONCLUSION

Our brains are complex information-processing systems, ones that are also *human* and *biological*. Thus, they manifest many quirks. Our brains "automate" as many functions as possible to speed up response time and to save energy by drawing upon past solutions. However, since our brains have only a limited supply of energy, and because biological systems are slower than computers, a reliance on "past solutions" helps to quicken our ability to respond to environmental challenges, which can be highly adaptive in a dangerous world.

Furthermore, our brains can only "think about" many events after meaning has been attributed to incoming sensory data. However, meaning fundamentally alters our perceptions by focusing on some attributes of an object while ignoring other attributes.

We have in this chapter considered two general principles that are rewritten here:

PRINCIPLE 4: THE AUTOMATIC PRINCIPLE

In comparison to a computer, our brains process information rather slowly. However, since our brains evolved in a very dangerous world, there was a need for quick action. Thus, to compensate for slow speed, our brains rely on automatic reactions as much as possible because automatic reactions are quicker. Automatic processes, though, often neglect important information in the name of expediency, with the result that our "thinking" is always biased and sometimes wrong.

PRINCIPLE 5: THE HUMAN FACTOR PRINCIPLE

One of the principal characteristics of being human is that we use symbols to interpret the world. However, by reducing events to symbols, be they words or numbers, we alter our perceptions and memories of the very events we are thinking about or remembering.

* * *

Up to this point, we have discussed five principles that help lay the foundation for other concepts yet to be presented. We have one more key principle to cover before getting into the specifics of memory, learning, emotion, and thinking. Now we will turn to the crucial topic of "development."

THE INDELIBLE STAMP
OF DEVELOPMENT

CHAPTER 5

We have already learned that our brains are extremely sensitive to context and different environments. Our brains begin life with extraordinary flexibility, partly to prepare them for the future. However, few concepts pose more difficulty to portray than brain and cognitive development. Capturing the brain's relationship to behavior is like the process of painting a picture of a three-month-old puppy—the darn dog will not stand still!

Throughout its evolution, the brain retained one very important characteristic—an ability to adapt and learn. Our brains change in response to the world. However, there are different types of change. Here we will focus on development: the unique transformation that results from both growth and some experiences. Developmental changes prepare our brains for future learning and thinking, indeed, for life itself.

Clearly, the structures and regions within our brains have been genetically determined. The brainstem, for example, is limited to certain types of tasks, it can never take over "thinking" if the neocortex is destroyed. Similarly, our brains have "language centers" that cannot relieve the brainstem of its duties. Nevertheless, within the cortex, different areas can be *enriched* when those areas are exposed to the right lessons at the right time, and the reverse is also true. The brain can be exposed to the wrong lessons during critical periods, with lifelong negative consequences. Shortly, we will consider what "right" or "wrong" might look like, but we need more background first.

Developmental processes are important for our understanding of both the brain and behavior, but these processes certainly complicate matters. If we want to understand memory, for example, we not only have to think about various types of memory, but we must also consider someone's age when a memory was formed.

As if our brains were not complicated enough, all areas of the brain do not develop at the same rate. "Lower" regions mature earlier than "upper" regions. By "lower," in this case, I am referring to the spinal cord, brainstem, and possibly some parts of the limbic system. "Upper" refers to the neocortex. This growth pattern is critical because basic life support functions are controlled by the brainstem, and if those functions were not operational at birth, we would not be here. Critical thinking, in contrast, which largely stems from the neocortex, is a skill that can wait to be acquired when we are older, as long as we are protected by adults during infancy and early childhood.

In addition to different rates of maturation from lower to higher, the brain *as a whole* goes through changes. For the sake of our discussion, I have divided these changes into stages. However, the following stages were arbitrarily chosen, and they are depicted as "stages" only because they describe processes that have beginning and ending points. Some processes span several years and terminate gradually, and some may show different patterns from one person to the next.

PRENATAL DEVELOPMENT

The first phase of an individual's development begins at the moment of conception. Among biologists and developmental psychologists, there is an apt phrase that sums up human development: "Ontogeny recapitulates phylogeny."[1] What exactly does this mean?

During prenatal growth, an embryo (and later, the fetus) passes through a process that appears to mirror the evolutionary pattern of the entire human species. That is, the growth of a single offspring *duplicates* the evolutionary history of the species; hence the adage that "ontogeny [the individual's course of development] recapitulates [concisely repeats] phylogeny [the evolutionary history of the species]." This idea may seem preposterous, until we look at the specifics.

We all begin as a fertilized egg (which represents the very earliest days of our evolution). After that, the egg divides. At about two weeks, nerve tissues form. By ten weeks, a single cell has become multicelled, and the process ends with billions of cells that look like a "complete human miniature" (the endpoint of our evolution).[2] Our individual growth "recapitulates" or mirrors the evolution of our species.

Aside from its interest value, there is a more practical reason why we should consider prenatal development. Early brain growth has enormous importance for our health. As we learned earlier, all the neurons we are ever going to have are generated prenatally during a process called neurogenesis. At a time when as many as fifty thousand cells are developing each second, the stakes are high. There is unparalleled potential for healthy development, but conversely, there is a great potential for abnormal development. *If* the developing nervous system is exposed during this period to harmful substances, like German measles or rubella, the developing person may be harmed for life. Other risk factors include nicotine, caffeine, or alcohol.[3]

Fetal alcohol syndrome (FAS) can result if alcohol is used during pregnancy, and this is an important issue because FAS is estimated to be the primary cause of mental retardation in the United States today.[4] Children with FAS may show permanent losses of intelligence, have facial deformities, or show behavioral problems as they grow older, such as lying, stealing, or problems with aggression and impulsiveness.

Fortunately, prenatal development unfolds without a snag for millions of children, laying the foundation for the next phase of development.

THE FIRST YEAR

At the moment of birth, another unique phase begins. Although the brain's neurons have been produced, the brain will still pass through many changes that are nothing, if not spectacular. Not to be upstaged by the two previous acts (evolution and prenatal development), the first year of postnatal growth will strongly influence the brain for the remainder of its existence. The first year is the brain's most important postnatal growth period for establishing healthy neurological foundations.

At birth, the brain is mostly *undeveloped* (even though most neurons have been laid down). Many of the connections between neurons and the dendrites of most neurons have not formed completely. This generalization does not apply equally to all areas of the brain, however, as the lower part of the brain, as we saw, is farther along at birth because it supports life itself. Nevertheless, as a whole, the new brain is only about one third the volume of an adult's brain, and during the first year after birth, the immature brain will double in volume.[5]

One of the more spectacular growth processes during childhood will begin during the first year of postnatal life. *Synaptogenesis* is the term applied to a rapid *increase* of synaptic connections within the cerebral cortex. The increase literally numbers in the trillions. It is this growth, in part, that accounts for the doubling of brain volume during the first year.

During the process of synaptogenesis, following a huge increase of extra connecting links, *many of those links will die off*! Eventually, the number of synaptic connections within the child's brain will *decrease* to the numbers found in adult brains.[6] This process *sounds* interesting, and perhaps is more intriguing than a good mystery novel, but what does it mean? Why would the brain grow so many extra connections, only to cut most of them back at a later date? The key to this mystery can be found by examining what happens to those connections that remain and those that disappear.

There is an *exact* correspondence between the initial growth spurt of synaptic connections and certain behaviors. Our first attempts to mimic a single word appear at about eight months of age, at the time when synaptic overproduction first begins, in an area of the neocortex associated with language. There is also an *exact* correspondence with other *types* of behavior at the time of maximum overproduction. At about eighteen months to two years of age, most of us were beginning to make simple sentences and our vocabularies were increasing.[7] This latter development occurs when the enormous volume of synaptic connections have reached a peak.

Researchers believe that the extraordinary number of connections, generated by synaptogenesis, is far too many for each connection to have been orchestrated by specific genetic directives. That is, too many synapses are generated for scientists to believe that each one is orchestrated by a genetic code. Rather, synaptogenesis, *as a whole*, is believed to be genetically mandated, but individual connections are not. The process may be difficult to conceptualize (it was for me), but an analogy might help.

Imagine a vacant lot overgrown with tall grass, which is well traveled by hundreds of little schoolchildren, decked out in their school clothes, carrying their little lunch boxes with pictures of dinosaurs on them. Initially, several different paths become worn through the grass, but some trails are inevitably more popular than others, so after a while, some trails are firmly beat into the ground, but most fade away. The same thing happens in the brain. The growth of synaptic connections is analogous to the growth of weeds in a lot that will later form the raw material from which paths will be worn.

Because one aspect of synaptogenesis involves the elimination of connections, after an initial increase, there appears to be an *optimal* number of connections that result in the most effective functioning. Too many connections may slow down or impede effective processing; too few connections may also impair processing in other ways.[8]

When certain neural pathways are used during development, those pathways form enduring connections that underlie *certain types* of learning. I said "certain types" because the process of synaptogenesis is believed to be the cause of "critical periods."[9]

Critical periods refer to times in our development when certain experiences must occur within a limited window of time, or the opportunity to "learn" may be lost forever. For example, children who are exposed to the *sounds* of any language early in life will learn to speak the language like a native. In contrast, adults who try to learn a new language *may* never sound like native speakers. Even if adults *do* master the sounds of a new language, they will never learn the language with the same degree of ease that unfolds naturally for children who are exposed to the language early in life (children with normal hearing and intelligence). "Sensitive period" is another term used to define a window of time during which some experience must take place, or the opportunity may be lost.[10] However, the term sensitive period is often used to refer to a window of time that is far more flexible than a "critical" period. For example, many people can learn to ski throughout their lives, but when children are exposed to the slopes during childhood, they often develop greater competence and athletic ability than those who pick up the sport at forty. (If sensitive periods show greater flexibility than critical periods, it may turn out that sensitive periods result from the brain's greater flexibility during childhood, as opposed to the process of synaptogenesis.)

In the realm of emotional development, if a child is exposed to a loving mother during the first year of life, normal attachment and bonding will occur. If children are removed from their primary caretakers during their first year of life, and bounced around from one caretaker to another, they may forever lose the capacity to bond with others.[11] Even death is possible if the ensuing neglect is pronounced.

I should add, while working on this project, I encountered a newspaper article that contradicted this theory.[12] According to the article, one research study did not find the mother-child bond (during the first year of life) to be predictive of later healthy adjustment. This finding was surprising because numerous psychological theories have long assumed that bonding during a child's first year is perhaps the most important mile-stone for healthy adjustment. Whether the study in question turns out to be valid or not, though, the existence of critical periods is well documented.

At this point, experts simply do not know how many experiences reshape parts of the brain during what time frames. Certainly, more is known about language in this regard than many other processes. It is likely, though, that humans do pass through *many* critical periods. For example, people who are born deaf but later have their hearing restored through surgery are unable to make sense of sounds, even when "normal" hearing is restored. The same holds true for normal visual development. Emotional development, too, is likely to depend on critical or sensitive periods, and there is clear evidence that parenting practices during early childhood produce profound effects on children. However, there is less agreement about whether specific practices must occur, and if so, experts are not certain about exact time frames.

During its lifespan, the brain shows two different patterns of change, one of which is most pronounced during childhood, as we have already seen. "Experience-expectant" development is the type of change that re-sults from synaptogenesis.[13]

Presumably as a result of evolution, the brain is believed to have developed a pattern that enables it to master those lessons that are re-quired by all members of the species. In effect, the brain "anticipates" the occurrence of *certain lessons*, and it does so by overproducing synapses. Different areas of the neocortex are prepared for the right learning at the right time. The example we are most familiar with is language. The brain "readies" itself for language by overproducing connections in those brain areas associated with language.

The types of changes that result from synaptogenesis (experience-expectant processes) appear to be in a class of their own. These changes *physically* restructure areas of the brain so it can accommodate future learning. The physical alterations appear to increase the volume and com-plexity of connections where the process unfolds.

Clearly, not all learning restructures our brain in the ways just dis-cussed. Not all learning is dependent on the process of synaptogenesis nor does it require a critical period. "Learning," as a general term, occurs throughout our lives, but synapse overproduction occurs only during critical time frames in childhood. Therefore, another process must account for "normal" learning. That process is called "experience-dependent" learning (in contrast to experience-expectant).[14]

To illustrate, I may have a special interest in jazz, so there are ways for my brain to accommodate my special interest (without the need for the entire species to undergo a critical period for jazz). It is now believed that the brain masters specific (idiosyncratic) lessons by modifying *some* synaptic connections (through the use of neurotransmitters) or generating a richer supply of synapses, but only in a specific area of the brain.

To clarify matters, we can compare the two types of changes we have just discussed. Experience-expectant processes involve phenomena that *all* humans are expected to encounter. All of us are "prepared" to begin language training by early childhood, so our brains are readied for language through the overproduction of synapses in key areas of the brain. Experience-dependent processes are those that allow individuals to learn specific lessons that are not needed for all humans, such as the ability to play the saxophone. Both of those processes involve synaptic changes. However, experience-expectant processes involve a huge increase of synaptic connections, followed by a whittling down of those connections. Experience-dependent processes either modify existing connections, or perhaps involve the *modest* increase of new connections, for a specific individual and purpose.

EARLY CHILDHOOD TO ADULTHOOD

Many of the processes we have been discussing so far begin during the first year of life, but extend well beyond the first years. Synaptogenesis is a case in point. It was placed under the category of first-year development because it *starts* during that time.

In addition to the marvels that commence during our first twelve months of postnatal life, our brains continue to change for many years in ways that are no less important for optimal functioning. During the second year of life, our brains undergo their next largest growth spurt.

In addition to increases in volume and changes in synapses, our brains undergo other transformations. Studies that measure the electrical activity of the brain have shown that the brain passes through five different periods of change during its first twenty-one years of life. These changes are termed "cerebral maturation" by the researchers who reported them, referring to when different parts of the brain become fully mature. Also, as if brain development were not complex enough, changes occur differently from the left cerebral hemisphere to the right one.[15]

One surprising finding of this research, in my view, is that the brain shows maturation until the age of twenty-one (on average). Twenty-one is older than what had commonly been depicted among the most prominent theories of development, like that of Piaget, relative to the time when the brain is still undergoing growth. However, the full ramification of this finding is not clear. Presumably, some twenty-year-olds may not be able to learn certain lessons that are accessible to fully mature twenty-one-year-olds. Just what those lessons might be, though, is not known. (Compared with studies relating to children, research on young adult development is practically nonexistent.)

There are yet other changes within the brain that are essential for optimal functioning. Specifically, changes in myelination occur until adolescence, or later. (Recall that myelination involves the process of covering some neurons with glial cells, which is analogous to covering the wires of an extension cord with a plastic coating.) Myelination increases the speed and efficiency of neural transmissions, which, in turn, improves mental functioning. Full myelination of the frontal regions of the cortex and the corpus callosum may not be complete until the age of twelve years.[16] (The corpus callosum is a bundle of nerve fibers that connect the two hemispheres of the brain.)

To give some idea of the importance of myelin, consider what happens when it fails to develop or is destroyed by disease. Multiple sclerosis (MS) results from the deterioration of myelin on motor (muscle) or sensory nerve pathways within the central nervous system. Individuals with MS might experience weakness in their limbs, lose voluntary control of their arms, or even become blind. The deterioration of myelin in multiple sclerosis is often patchy, so a lost function may return. Symptoms of MS may therefore be short-lived or lifelong.[17]

As explained, different areas of the brain develop during different time periods. Lower areas of the brain mature first, upper areas mature later, often into adulthood. This is an important concept because it also means that different areas of the brain are more susceptible to influences, good or bad, depending on age.[18] Thus, what might be harmful to a three-month-old baby may have little impact on a fifteen-year-old teenager.

Without doubt, developmental changes result in the types of changes that permanently alter the brain and prepare it for later life. These changes exert their influence for the remainder of our lives, and this general idea is encapsulated in the next of our general principles.

PRINCIPLE 6: THE DEVELOPMENTAL PRINCIPLE

Experience and growth during the first few years of life physically alter our brains and prepare them for future learning, thinking, and behavior.

The many changes that occur to our brains, both prenatally and during the first years after birth, underscore the importance of childhood experience, because those experiences prepare us for the future. However, there are two corollaries to this general principle.

First, exposing children *to the right lessons* will have lifelong positive consequences—a fact that has been known for years. In this instance, the right environment includes age-appropriate and safe objects for play (including objects that provide a rich variety of colors, textures, sounds, and objects that can be safely manipulated, or even put into the mouth). *Most important*, though, the right environment involves interactions with loving adult caretakers. Such an environment must also be safe, free from violence or threats of violence, and it must be free of other dangers (such as toxic substances, including lead, cigarette smoke, and the like).

Second, unfortunately, is a reverse scenario. Exposure to negative environmental influences, such as violence, neglect, or substance abuse, can also result in permanent changes to the brain; and some negative outcomes cannot be eliminated by later attempts to ameliorate the damage. Fetal alcohol syndrome provides a good example. Although the effects of FAS can be mitigated to some extent, the deleterious effects cannot be completely erased through any known means.

Negative influences embody two possibilities where the brain is concerned. First, deleterious outcomes may occur when the brain does not receive the types of experience and exposure that is needed for optimal growth during a critical period. If a child is placed into an orphanage at birth and given very little human affection or contact, brain size can be reduced and healthy emotional development may be thwarted.[19]

Less than optimal brain development may also occur in another way. In some instances, children may achieve normal brain growth, but their brains might still have been exposed to lessons that were highly inappropriate. Consider what would happen if all the language lessons to which young children were exposed involved only crude or sexually explicit words. The brain would still develop in response to those words, but the child's resulting behavior would be *abnormal*. Children can be very intelligent, but simultaneously be exposed to certain influences that have devastating, lifelong consequences.

As a psychotherapist, I have met many adults who were extremely intelligent, but who had been sexually abused as children. Generally, when children are exposed to developmentally inappropriate lessons, like sexual abuse, those children often develop mental and behavioral patterns that create lifelong problems for them.

As adults, sexual abuse survivors often have problems with closeness in relationships, or may misread social cues. For example, if a sexual abuse survivor's partner wants to be physically close, the survivor might misread the cue, thinking that a request is being made for sex. In general, issues relating to nonsexual intimacy and intimate sexual relationships are often muddled and difficult for sexual abuse survivors. I wish to emphasize, though, these individuals may have normal intelligence, but their early experiences have left them with perspectives about relationships (or other people) that continually influence their reactions within a social world. (In contrast, children who were severely neglected may end up with brains that are actually smaller and be less intelligent.)

The idea that the brain is changed through interactions with the environment was popularized by the Swiss biologist, Jean Piaget.[20] He achieved fame by describing several stages of cognitive development that were hypothesized to unfold in a preordained order for all children of normal intelligence—and just about every adult since Piaget has observed the same phenomenon in small children.

All children, no matter how intelligent, before age four or five, are unable to distinguish "reality" from appearance.[21] For example, if a person dons a Santa Claus costume in full view of a three-year-old, in the eyes of the child, Santa magically appears. However, once children undergo a "developmental leap," which is presumed to reflect actual brain growth, they readily understand the illusion created by a costume.

For the most part, theories of human development have focused on the changing abilities of children. That has always seemed appropriate because growth and development are most pronounced during childhood. However, adults, too, show differences from one another relative to their intellectual abilities. There is no universal agreement, though, about whether some of these differences are reflections of underlying brain development or just differences in past learning and education.

"Higher" (presumably more sophisticated) stages of cognitive development have been hypothesized to exist for some adults. One such stage has been termed "post-formal operations."[22] This style of thinking is unique to humans, but presumably it is only found among *some* humans. Thus, even individuals with average intelligence may not develop post-formal operational thinking.

Post-formal operational thought describes a thinking style that permits us to recognize that even our most cherished beliefs *might* be mistaken, or that two diverging points of view (that appear to contradict each other) *might* be equally valid. How could that be? One of the top political issues of our day pertains to abortion. One individual might believe that abortion is murder, while another might believe that abortion is *not* murder. If both individuals can maintain their own views and still acknowledge that other points of view *might* be valid, that ability is post-formal operational. This style of thought is highly abstract and considers the contextual nature of many truths.

Researchers, as noted, have spent relatively little time considering adult development in comparison with the amount of time focused on child development, so the implications of a higher level of cognitive development are not clear. Certainly, there is nothing that approximates universal agreement that such a "stage" even exists. However, some of what researchers have discovered about children's development can be applied to adults as well.

Studies on thinking, for example, can be generalized to adults and children alike. Effective and less effective thinkers, for example, have been studied, and differences between these two groups have been identified. The best thinkers spend more time contemplating their own problem-solving strategies. In turn, when we contemplate our own mental processes, we are more likely to spot problems in logic. Spotting logical errors, in turn, often produces a more thoughtful exploration of alternatives.[23] The reason that differences in thinking superiority are raised here is because one of the hypothesized characteristics of post-formal operational thought is a difference in thinking superiority.

We learned earlier that the brain's first response is *not* to second-guess its own conclusions; rather, its first response is to believe. It would appear, though, that more gifted thinkers recognize the possibility of error and are more likely to look for error. Clearly, some people are more adept at seeing numerous aspects of a problem, while others are stuck with a single perception. Part of the reason for this difference *may* stem from differences in brain development. One reason for this speculation is that adults often show differences in thinking, contrary to a common myth.

It is often, but *erroneously*, assumed that when a certain age is reached, people are automatically endowed with certain *minimal* abilities. This assumption, though, may be incorrect. Why?

None of us has achieved the same level of competence or knowledge in every single area of knowledge. A baseball enthusiast does not have the same experiential understanding of baseball as a professional player; a chemistry professor has a far greater understanding of scientific methodology than a beginning chemistry student; an adult who has spent a lifetime studying philosophy has a better understanding of moral issues than the general prison population. The foregoing differences *may* be caused by differences in education or experience, but as we have seen, distinct experiences may endow each of us with different levels of development, *relative to a specific area of knowledge or expertise*. Even if the differences we see among adults are simply differences in education (and not underlying brain development), the result is the same. Those with a less sophisticated level of understanding, *in a specific area*, are often blind to areas of knowledge that have not been achieved. Again, this is an important issue because adults are often thought to have minimal levels of understanding, by virtue of having reached adulthood, and that may not be the case.

When an individual is incapable of seeing an argument or idea that another can readily see, I refer to this as a "developmental blind spot." We can readily see these in children by looking at their reasoning abilities. The very young have a different way of looking at the world—and no amount of instruction can raise their level of awareness to that of older children until actual brain growth has occurred. A small child might believe in Santa Claus, not because Santa exists, but because a three-year-old can believe in Santa Claus, whereas a fifteen-year-old of normal intelligence cannot swallow the notion of a fat man sliding down someone's chimney.

Adults, too, may show differences in development, but the idea of adult development is difficult for some people to believe. Even the most disturbed adults often are incredulous at the suggestion that there *might be* matters pertaining to their own behavior and mental processes about which they have little awareness. Nevertheless, like a three-year-old who thinks he is invisible when he hides his eyes, most of us manifest "developmentally immature" beliefs from time to time (and those beliefs guide our thinking and behavior, often unfolding outside of awareness).

An individual's lack of awareness about a level of development that has not been achieved is one of the major tools used in psychological testing. I once performed a psychological evaluation on a woman for a custody case. Her children had been removed from her home because of allegations of severe neglect. Testing was requested by social services to determine her ability to learn information about parenting. In talking with her, it became obvious that she wanted her children back, so motivation was not an issue. During the testing process, though, she attempted to "fake" two psychological tests by making herself look as if she had *never* had any problems. Her lack of sophistication was quite apparent. She demonstrated a very simplistic notion about mental health, and her perspective was easy to spot. In order to check my theory about her, I spent a considerable amount of time interviewing her. At one point, I asked her what it *might* have been like for her children to have been removed from their home and placed in foster care. Most parents, in my experience, would have some opinion about the feelings of their own children. "It was terrible for Johnny." "Little Sheila was so frightened." "Albert was so mad." "I feel guilty because I caused this." In the case of this woman, though, her lack of empathy for her children was very pronounced. Even after several visits with her children, she still had no idea about their feelings or an opinion about how their removal from the home might have affected them. (I should add, so I do not leave the wrong impression, I still recommended *strongly* that the family be reunited as quickly as possible and be provided with whatever help might be needed to allow them to function more optimally. The goal in this case was to assist the woman in learning how to take better care of her children.)

The fact that different individuals of similar intelligence often show differences in how they perceive or process information is important for understanding humans and for understanding conflicts that arise among them. For example, when a religious book is being discussed, one adult might interpret the text literally and concretely, while another might view the book allegorically. Their disagreement is often a genuine difference of opinion, but it can also be caused by *dissimilar* processing styles. Some individuals may be *unable* to consider the material abstractly.

A difference in development or cognitive processing style, because of developmental factors, is an important social issue because the process of debate cannot influence the opinions of people who are *unable* to see another perspective. Thus, what might appear to be a debate may not be one at all. If some humans cannot comprehend an opposing opinion, then an argument is not being judged on the merits of the argument; rather the outcome is being decided by the fact that an individual cannot comprehend one side of the "debate." (Although this is an important point, it is not without its critics, and as you can see, has potential to be a highly contentious issue.)

Before leaving the wrong impression, I should also emphasize that we often show a bias for higher levels of functioning. In any given field, "smarter" is often viewed as "superior." However, lower levels of development on a given topic are not necessarily inferior, from an *adaptive* perspective. For example, it is probably easier for the average adult to explain why punishment for a crime is appropriate. Generally, it is more difficult to explain the merits of "forgiveness" or "rehabilitation." We can test this by asking our friends what they think. (I am not saying that one approach, punishment or forgiveness, is better than the other, only that one is *easier* for the average person to understand and justify.)

From a developmental perspective, the "punishment" argument is developmentally less sophisticated, insofar as it requires less abstract reasoning ability. From an *adaptive* perspective, though, the death penalty might be more functional.

It is probably safe to conclude that different levels or styles of thinking are appropriate for different circumstances. Also, it is very likely that *all* adults manifest lower (and higher) levels of cognitive or moral development at some time or another. Under extreme duress, each of us might resort to the logic typical of a small child. None of us are immune to panicking during an emergency, for example, and during those times, we often draw upon childlike solutions.

With little doubt, we often rely on beliefs that formed early in our lives and perhaps show no more sophistication than what might be expected from a six-year-old. Further, because the brain requires a constant supply of food and oxygen, and because more in-depth thinking requires more fuel, greater analysis occurs less often.

To place our discussion in a broader context, two additional ideas must be added to the mix. First, higher levels of development or sophistication can only be considered higher in light of a *specific domain*, like numerical reasoning. Even the most gifted individuals, like Einstein, are unlikely to have been uniformly good at all things. In spite of many crowning achievements, I am sure that even Einstein had some areas that were less superior outside specific areas of superiority.

Second, the fact that an individual has reached a certain age does not guarantee that even minimal skills or abilities exist, nor does a certain age guarantee that certain abilities can ever be achieved.

Many years ago, I worked with a thirty-three-year-old father of two who was married, fully employed, and quite bright. One day he commented on a newspaper article about a grave robbing. To this day, I have never forgotten the gist of what he said. He believed that grave robbing should be a capital offense. According to him, kidnappers often receive the death penalty, yet their victims start out being alive, so their victims have more to say about what transpires; but in the case of a buried body, the body has no influence over what happens to it. Therefore, grave robbing is worse than kidnapping. (At the time, I could not respond to his words, but intuitively, I felt there was something questionable about his logic.)

It is not uncommon to see examples of logic that appear flawed, yet have been produced by individuals who appear intelligent. Clearly, critical thinking can be learned, and probably must be taught (which is one reason people often spend many years in school). However, what is not known, at present, is how much of an adult's ability to think abstractly is based on brain development and how much is based on learning and education.

We have focused on some of the many changes through which our brains pass on the road to maturity, and we particularly concentrated on differences between individuals that *might* be attributable to development. Most of the processes mentioned here merit entire textbooks. This overview, of necessity, has been brief and provided for illustration only to add enough background for Principle 6 presented earlier.

I wish to emphasize that the linkage between cause and effect is not established in all the examples we will be looking at, but the malleable quality of the brain during its early years of life is well established.

IMPLICATIONS FOR UNDERSTANDING BEHAVIOR

Functional Impairments

During a workshop I attended in 1996, the presenter showed a photograph of a brain scan that he had previously shown to a colleague. His associate thought the scan depicted an adult with Alzheimer's disease, judging from the brain's shrinkage. In fact, the scan showed the brain of a seven-year-old girl who had been neglected. Simple neglect, according to the expert, resulted in a smaller brain, which also caused an intellectual loss equal to twenty or twenty-five IQ points.[24] In practical terms, that is the difference between average intelligence and mild mental retardation. A few months after I attended the workshop, I encountered two individuals in my professional capacity as a psychologist who, on the face of it, appeared to lend substantiation to what I had learned.

Over the course of my professional career, I have administered hundreds of intelligence tests. Usually, I can make a ballpark estimate of intelligence after an extended clinical interview. A clinical interview is a systematic way of getting background information, to include such things as place of birth, education, family size, scope of the problem, and so on. Usually, when someone is mildly mentally retarded, that fact is often apparent during an interview. During a two-year period, though, two patients completely thwarted my powers of prediction. I fully expected each individual to function in the average range of intelligence, but I was wrong. Each person manifested a certain "brightness" and social awareness that made each one seem intellectually normal. As it turned out, much to my surprise, both individuals scored in the "mildly retarded" range. (IQ tests have serious limitations, without question, but they generally do an adequate job at measuring some general intellectual abilities that are useful in this society.) I later learned that both individuals had long histories of neglect and sexual abuse. I am now convinced that both individuals were born into the world with normal brains that failed to develop fully because of early childhood neglect and abuse. Naturally, I cannot prove this theory, but the linkage between neglect and intellectual impairment is growing.

Recently, I came across a research article entitled, "New Zealand Study Lasted 21 Years."[25] The study was described as the "most comprehensive" of its type. At the beginning of the study, one thousand babies were selected to participate. All were born in 1972 and were examined and interviewed every two years. (Starting at the age of fifteen, study participants were only reevaluated every three years.) The families of the one thousand study participants were also interviewed on the same time schedule.

One of the major findings was that, *by three years of age*, children with "serious problems" (described as poor concentration or overactivity) had a 30 percent higher risk for delinquency.

The synopsis I read did not show what caused certain problems, like poor concentration, but it speaks to the general issue that early problems in childhood often have lifelong influences, a fact that is well documented and worth heeding. However, the information presented earlier in this chapter suggests a rationale why earlier experience might have lifelong effects.

Through the process of synaptogenesis, a child's experience becomes permanently imbedded within the brain. Thus, a child with poor concentration establishes a poor foundation on which to build future learning. Since the brain is highly dependent on each experience, if a child does not pay attention, he would be failing to learn "how to pay attention." His brain, in effect, would be failing to learn "how to learn." Thereafter, the stage has been set for later school failure. Once school failure has occurred, the chances of delinquency rise dramatically.

Adult Children

All individuals who provide professional psychotherapy services are familiar with the term "personality disorder." This is a general class of mental disorders, ones that encompass abnormal conditions that are first apparent in childhood or adolescence.

By definition, an individual under the age of eighteen cannot be diagnosed with a personality disorder unless the impairment has lasted at least one year. (There is one exception for a specific disorder.)[26]

Personality disorders are described as "enduring" patterns of behavior that show substantial deviation from behaviors that are culturally appropriate. Individuals with personality disorders are often embroiled in tumultuous relationships. They often have difficulty in problem solving, emotional expressiveness, or problems controlling impulsiveness, temper, or both. The problems faced by these individuals (and those around them) exact high costs in the social arena, affecting work and personal relationships. Finally, these individuals often experience legal difficulties.

Although each type of personality disorder is described in a book about mental disorders, personality disorders are not considered to be a disease, like schizophrenia. Rather, there has long been a consensus among experts that personality disorders are the result of negative environmental influences, like childhood neglect or abuse or exposure to criminal activity in the home when a child is at an impressionable age.

Interestingly, recent brain-imaging studies have shown abnormal brain patterns in some individuals with personality disorders.[27] What is relevant here is that the types of abnormal development described by neurologists are consistent with what we would predict for children who were either abused or neglected or spent too little time in the presence of adults while they were growing up. In others words, it may well be that personality disorders are caused, in part, by abuse and neglect, which, in turn, reshapes the brain, particularly the frontal lobes. (While speculative, there is no good evidence to dismiss this theory out of hand.)

What is most striking about individuals with personality disorder diagnoses is the degree to which they *act* (in specific situations) like young children. Some are unable to control their temper. Some reek havoc on the lives of others through stormy interpersonal relationships. Some are sexually promiscuous. Some are completely impervious to the effects they have on others. Some engage in risk-taking behavior to the point of being dangerous to themselves or others.

The fact that personality disorders show symptoms in childhood or adolescence speaks to the probable influence of early developmental factors. Generally, their symptoms are not treatable through medication (although they often have other symptoms, like depression, which can be treated by medication). The best working hypothesis to explain at least some of these problems pertains to brain development. Specifically, the brain's capacity to be permanently altered by early childhood influences.

Imagine someone of normal intelligence who *cannot* learn, at least in a circumscribed area of functioning. If you can imagine this, or perhaps know someone who fits this description, then you have a good idea of what these individuals are like from a professional's point of view.

I once treated a man who had no close friends because, to him, relationships were obstacles. Every week he would come in and talk about what others had done to him, how they had "done him wrong," "how he was *always* in the right." I never heard him admit to any responsibility for his problems, but at the same time, I believe that he was accurately describing his perception of the world. Virtually no amount of interaction on our part could or would budge his way of looking at the world, and his behavior remained intractable for months on end.

This man believed that others were not as good as he was, so he was continuously taken aback when others did not realize how "superior" he was (in his eyes, at least). When others failed to see how talented he was, he would seethe with anger, yet rarely show his rage. Rather, he would secretly fantasize about getting even (which he never did, to my knowledge), although when he was younger, he often walked off jobs in a fit of anger and never returned. The man, in his late sixties, had shown the foregoing pattern since his early teenage years.

Finally, if drugs or alcohol are being abused by these types of individuals, we can end up with some very dangerous individuals. The reason is that alcohol and many drugs diminish the ability to control impulses and behavior, and among individuals who are already impaired in these areas, a further reduction in control can be disastrous. Thus, substance abuse treatment is often necessary, in conjunction with other forms of treatment.

At present, most authorities are divided on the matter of treatment for individuals with personality disorders. One camp believes that these disorders cannot be effectively treated by any means. Another group of clinicians support the need for therapy, but *several years* of treatment (or even lifelong therapy) might be necessary, although the results of long-term therapy is by no means consistent among all patients.

My own view is that outpatient treatment, through mental health centers or hospitals, can be very helpful, but mostly for providing external controls and monitoring to help them function in the community. The alternative is jail or prison or long-term hospitalization, all of which can be quite expensive (not to mention inhumane for individuals with mental disorders). The issue of how to handle these people within society is not an easy one. Neither lifelong mental health care nor locked settings, like jails, come inexpensively, and different settings have both advantages and drawbacks.

Substance Abuse

The issue of substance abuse will be revisited here because of its relevance to the developing brain. As noted, the brain is still showing growth changes, possibly as late as twenty-one years of age. Alcohol, nicotine, and other toxins can impact brain growth both during and after prenatal development, and these toxins can harm the brain in a variety of ways.

As discussed, alcohol can cause physical deformities or mental retardation, or both, especially when it is ingested during pregnancy by the mother. Other substances, such as lead, cocaine, and the list goes on, can also affect the developing brain. This point, while important, will not be explored further here because most of us have heard about the *direct* effects that toxins can have on the developing brain. But there are yet other ways in which alcohol and other drugs can cause harm to the developing brain.

If children or adolescents are out drinking somewhere, then the time spent drinking is time away from other activities that *could* produce beneficial results for the growing brain. Assuming that the brain is developing until twenty-one, one assumption is that early learning is restructuring the brain for later life, so the brain needs to be challenged with the right lessons during these formative years. If kids are drinking or experimenting with drugs, then the experimentation is unlikely to be the best environment for the developing brain.

During a time when the young brain is particularly susceptible to learning, substance use may be preparing the brain in the wrong way. The process of chemical "addiction" occurs when brain cells "learn" to expect a chemical that has been taken before, and during childhood and adolescence the brain is even more sensitive to these learning influences. Thus, adolescent experimentation with drugs or alcohol may be particularly risky, in terms of setting the stage for future addiction.[28]

There is yet another way in which alcohol and other drugs harm the developing brain during adolescence. Adolescence is the time of development when teenagers *should* be learning more and more about how to relate to others, how to control impulses, how to empathize with the feelings of others, and how to "think through" problems. Generally, it is a time to learn to communicate more effectively within a social arena (outside the home).

Nevertheless, substances like alcohol and marijuana alter the brain's ways of perceiving and processing data. The output of the brain under the influence of chemicals is effectively altered. Perceptions are altered, emotions are altered, and thinking is influenced. Thus, during the time when the brain is most ready to learn about the social world, its normal functioning is being chemically altered in unpredictable ways. Thinking, arising from the cerebral cortex, learning, and even emotional function all provide a type of information about our relationship to the social and physical environment that cannot be obtained in other ways. Most drugs, though, including alcohol, can alter cognitive and emotional processes in unpredictable ways, "unpredictable" because different doses and different drugs have varying effects on different people. In sum, children or adolescents may be learning about the social world on the basis of inaccurate information, that is, from the viewpoint of a brain that has been chemically altered. If, however, their perceptions have been chemically altered, they may miss out on important learning.

Prevention, Prevention, Prevention

Where critical and sensitive periods are concerned, timing is everything. Thus, if we want our children to speak a foreign language like a native, sooner is better, but certainly during early childhood, if possible. Spending time with children, holding them, reading to them, exposing them to music, and providing them with stimulating, but safe objects are all ways of helping a child's brain develop in a healthy manner.

The prevention field involves many specific areas. There is substance abuse prevention, violence prevention, disease prevention, HIV prevention, and pregnancy prevention. If there is a problem, it can be prevented before it occurs.

Years ago, when I first began working in the mental health field, prevention efforts often involved treating problems *after* problems became apparent. (Although treatment is important, it does not *prevent* problems before they occur.)

In contrast to treatment, one of the more promising developments in the field of substance abuse and violence prevention involves the study and identification of "protective factors."[29] Protective factors are those strengths or attributes that appear to protect children from later acts of violence or drug use. Protective factors include close family ties, involvement in school activities, and a commitment to school success, to name but a few. "Risk factors," in contrast, which are those attributes that predict which children might need help, include frequent truancies, constant fighting, strained family relationships, and so on. When potential risk factors are identified early, more effective attempts can be made to intervene before more serious problems arise.

The prevention field has come of age, and both research and intervention strategies are becoming available that prevent many problems before they occur.

CONCLUSION

We have already learned that during most of human history, *all* individuals lived in small hunting and gathering bands. Bands, in turn, consisted of small, extended kinship groups, numbering between twenty-five to sixty individuals. From a developmental perspective, it is inconceivable that prehistoric children were exposed to influences outside their bands. Although today's human beings are no less tied to their environments for survival, we moderns tend to alter our natural and social environments, often with little knowledge about the effects of those changes. The reason I reiterate the issue of social environment here is because the brain did not, and does not, develop or function in a vacuum. No other known organ is more responsive to its surroundings. Our brains develop in response to the world, and, in turn, that world reshapes the developing brain to accommodate future learning within a specific type of environment.

Most of what we have discussed here has focused on relatively permanent changes within the brain, and some of this discussion has emphasized the potential for deleterious consequences. Thankfully, however, millions of children, each year, are exposed to loving, nurturing influences, so those children are the recipients of positive and permanent changes.

The major lesson to be learned is that many experiences leave an indelible mark on our brains that no amount of later education, training, learning, experience, or even punishment can mitigate, and these themes are encapsulated in the Developmental Principle.

PRINCIPLE 6: THE DEVELOPMENTAL PRINCIPLE

Experience and growth during the first few years of life physically alter our brains and prepare them for future learning, thinking, and behavior.

* * *

We have now finished six general principles that form the foundation for the remainder of this book. In subsequent chapters, we will discuss specific functions of the brain, like "learning" and "thinking."

Although this discussion of development may have left the impression that we cannot change once our brains have reached maturity, that conclusion is not entirely accurate. We cannot change past experiences, many of which leave indelible marks on our brains. However, early development prepares our brains to live in the world, and the best-prepared brains have an extraordinary capacity to memorize, learn, and think.

SECTION I: SUMMARY

Up to now, we have considered six general principles. Individually, each principle may be helpful for understanding some aspects of behavior, but when all are reviewed together, we may gain a greater understanding. We are reiterating here, because all of them combined pertain to the remaining topics to be considered. "Memory," which is the next topic of discussion, should be considered in light of all six principles, and the same holds true for the subsequent topics.

Here are the first six principles:

PRINCIPLE 1: THE BRAIN GROWTH PRINCIPLE

The brain of modern humans reached its large size long before humans began to grow really inventive and create such things as written language, art, agriculture, steel tools, and large cities.

PRINCIPLE 2: THE SOCIAL ENVIRONMENT PRINCIPLE

The brain's most recent growth or evolutionary change occurred during a time when our ancestors lived in small groups or bands of twenty-five to sixty individuals.

PRINCIPLE 3: THE NETWORKING PRINCIPLE

Our brains are highly complex information-processing systems or networks that often segregate information. The concept of a "network" has two implications. First, for most behaviors and mental activities, such as thinking, many parts of our brains are temporarily linked together by brain areas called "association" and "executive" regions. Second, our brains are made up of many parts, some of which evolved during different periods of evolutionary history.

PRINCIPLE 4: THE AUTOMATIC PRINCIPLE

In comparison to a computer, our brains process information rather slowly. However, since our brains evolved in a very dangerous world, there was a need for quick action. Thus, to compensate for slow speed, our brains rely on automatic reactions as much as possible because automatic reactions are quicker. Automatic processes, though, often neglect important information in the name of expediency, with the result that our "thinking" is always biased and sometimes wrong.

PRINCIPLE 5: THE HUMAN FACTOR PRINCIPLE

One of the principal characteristics of being human is that we use symbols to interpret the world. However, by reducing events to symbols, be they words or numbers, we alter our perceptions and memories of the very events we are thinking about or remembering.

PRINCIPLE 6: THE DEVELOPMENTAL PRINCIPLE

Experience and growth during the first few years of life physically alter our brains and prepare them for future learning, thinking, and behavior.

* * *

From this point on, specific *functions* of the brain will be discussed, including memory, learning, thinking, and emotion. We will also be revisiting the issue of human evolution and address some general social concerns along the way.

MEMORIES: LOYAL AND MISUNDERSTOOD

CHAPTER 6

Perhaps no other mental phenomenon has been the subject of more misunderstanding than memory. *The American Heritage Dictionary* defines memory as "the mental capacity to retain and recall past experience."[1] As we can see, there is the assertion that memories can be recalled and that they are records of past events. Under some conditions, both of these conclusions can be *false*, and we will learn why.

There is probably no subject more intriguing, nor pivotal for understanding behavior, than "memory." We humans, perhaps more than any other animal, can devote our lives, for better or worse, to the service of our memories. Once past experiences have been encoded in our brains as "memories," we may behave forever afterward on the basis of those memories. However, since the original event has ended, are we behaving on the basis of the event or the memory? If we behave on the basis of a memory, to what extent can we disentangle ourselves from the past?

If we think about "memory" as a general term, it provides a good example of the brain's evolutionary growth pattern. Complementary functions are thought to have evolved during different time periods, presumably in response to different environmental demands. Gradually, each new function or adaptation provided a slightly different advantage, with the result that we have many memory systems.

Different types of memories distinguish themselves in a number of different ways. Further, the interpretation of some memories can change over time. Finally, stored information influences how we see the world, so *future* memory formation often is affected by existing memories. If all of this sounds confusing, it is—but this chapter will untangle the web.

At the most basic level, scientists have learned that even individual neurons form memories of sorts.[2] Relatively enduring changes that occur within the brain as a result of experience is a good definition of memory; therefore, we can conclude that neurons form memories. This is not to say, though, that individual cells contain information, as such. Rather, when cells are used, or used often enough, their dendrites and synapses may be altered.

Memory, as you and I experience it, is supported by thousands or millions of individual cells, working in concert, but at present, no one knows exactly how changes at the level of individual cells result in the phenomenon we experience as "memory." We can use an analogy, though, to conceptualize the process. Perhaps you have seen the desert after water has run over the ground. The water carves small streams into the ground, and those marks remain even after the water has evaporated. The ground remains altered. This analogy illustrates how memories might be formed. The stream bed *is* the memory.

As a psychologist, I have frequently heard professionals from various fields talk about "memory" as if it were a big storage closet in which everything is thrown together. Not long ago, I supervised therapists who often talked about their patients' memories, but they rarely tried to differentiate one type of memory from another. In fairness to my colleagues, though, it has been only since the 1980s that researchers have tried to develop language that more precisely describes different memory processes.[3]

Scientists now have identified many types of memory, but they have failed to devise terminology that can muster 100 percent agreement among all experts. In a perfect world, researchers would love to know how each type of memory is produced and exactly where it is stored within the brain. In that case, researchers would describe memories by designating their exact locations within the brain. To date, though, scientists have isolated only a few memory pathways.[4]

In the absence of full knowledge about specific memory locations and functions, scientists nevertheless must communicate with one another about various memory processes, so they have assigned labels that describe different types of memory. Thus, different terminology refers to the ways in which different memories can be distinguished from one another. I should add, not all researchers are pleased with this labeling system because it fails to communicate specific information about possible brain mechanisms. With these labeling shortcomings in mind, we will examine some different types of memory.

WORKING MEMORY

We all need what has been called "working memory."[5] It consists of the ability to hold something in consciousness while we are using that "something," but afterward, we may be unable to remember the information no matter how hard we try. I can remember a new phone number long enough to run across the room and jot it down, but beyond that, the number is gone from memory. Working memory is thought to stem from different areas within the brain, but different working memory tasks call upon different areas of the brain.[6]

For many years, scientists had hoped to isolate the exact location of working memory, but as we just saw, they have since learned that different tasks draw upon widely dispersed areas within the brain, sometimes cutting across hemispheres, sometimes cutting across the "evolutionary" boundaries we discussed earlier.

There are two components of working memory. We can temporarily store either visual, spatial, or auditory information and perform problems with that data. Thus, working memory is thought to consist of a "buffer" that holds information temporarily, while an "executive" region uses that information to solve problems.[7]

INTERMEDIATE-TERM MEMORY

Contrary to popular belief, there is not a direct transfer from short-term to long-term memory. Generally, short-term (working) memories completely disappear before long-term memory "traces" have been established (at least for some types of memories). Research with animals has shown that it takes several hours before memory consolidation occurs after a learning episode, and this fact is important because of the similarities between human and animal brains relative to many biochemical processes.[8] Further, rapid eye movement (REM), which refers to a specific stage of sleep, has also been implicated in the long-term memory consolidation process.[9] In one experiment, individuals who were deprived of REM sleep could not recall a task that was presented immediately prior to the sleep period.

Finally, and very importantly, studies with animals have shown that high levels of stress may interfere with memory formation and may even destroy neurons within the hippocampal formation, which is a key area of the limbic system associated with some forms of learning and memory. This finding is all the more ominous in light of what we learned earlier about brain development. This information, therefore, underscores the need for environments that are safe and relatively free from stress.[10] In short, at least some types of long-term memory consolidation processes are complex biochemical events that can take several hours to occur.

The issue of intermediate-term memory is of paramount importance to professional therapists. Since there may be a considerable time lag, possibly several hours, between the occurrence of an event and the formation of a long-term memory, a pertinent question emerges. Is it possible for a memory to form, but in an altered version from what actually transpired? This *could* be possible because there is evidence that memories *can* change in meaning over time.

LONG-TERM MEMORY

By current scientific consensus and for the sake of discussion, *long-term memory* has been divided into two general categories. As we have discovered, there is not a direct correspondence between the labels that are used to describe various memory processes and specific brain regions. However, so as not to be misleading, *some* important brain "parts" *have* been identified that are essential for some types of memory. Overall, though, entire long-term memory systems have not been fully identified or mapped.

The two general types of long-term memory include (1) *declarative* (or explicit) memory and (2) *nondeclarative* (or implicit) memory. Each of these systems stores different types of information, and each shows different patterns related to recall and the inability to recall.[11]

Declarative Memory

The terms "declarative" and "explicit" are synonymous. Both are used to denote a general *type* of long-term memory. Within the general category of declarative memory, there are many different types, including spatial, semantic, and episodic memories.

Spatial memories are consciously recallable memories about location. We can remember the way to the local mall and even describe the way to a friend who is visiting from out of town. *Semantic memories* are general facts and knowledge. We can recall our names, the names of different towns, even our dates of birth. *Episodic memories* refer to consciously retrievable knowledge about an event that occurred at a specific time, such as Aunt Millie's ninetieth birthday party.

Generally speaking, declarative memories are the quintessential type of memory. They are the types of memory that we normally equate with the term "memory." They are consciously retrievable and we can summon them on demand (usually). In short, stored facts and information that we can quickly access are *declarative memories*.

Retrieval and Forgetting.   As we can tell from experience, the retrieval of declarative memories is *usually* very quick. Normally, we do not spend much time retrieving declarative memories. When someone asks us what we do for a living, the information rolls off our tongues without hesitation. However, we have all experienced situations where we knew something, but temporarily forgot, just when we tried to retrieve the information. Like the bric-a-brac at the front of a hall closet, declarative memories are readily available—*if* we can quickly find the key. Unfortunately, we are stuck with figurative labels like "key" to describe memory processes. No one knows, or has been able to adequately describe, what we do *exactly* inside our heads that allow us to call forth memories. Although brain-imaging studies shed light on some areas of the brain involved in retrieval, those studies tell us little about *how* words are retrieved. What we voluntarily do to call up a piece of stored information is a big mystery. With declarative memories, though, we at least *know* that we are accessing memory (even when we fail to recall the word we are after).

Speaking of forgetting, declarative memories often fall prey to *temporary retrieval failure*, but there is enormous variability on this dimension from one person to the next. Declarative memories can also be lost forever, in spite of the once-common belief to the contrary. I recall an incident from several years back when an attorney asserted that memories could never be lost. That was, apparently, once a common belief, and even today, it is not uncommon to see a television program where, under the influence of hypnosis, someone remembers even the minutest detail from an event many years ago.

Memories are now believed to be formed by well-used pathways within the brain.[12] That is, when synaptic connections are used, memories form, presumably because of changes in the synapse. Conversely, when paths are not used often enough, memories can disappear. What complicates matters is that there are different types of memories, and some disappear more readily than others. However, there is no reason to believe that declarative memories cannot be lost. When they are lost, there are two likely reasons. First, synaptic connections can deteriorate if a memory is not used often enough. This is called the "decay" theory.[13] Second, memories can be lost as a result of brain injury or disease when neurons are destroyed.

You may recall times in high school or college where you crammed for a test, perhaps did well, and promptly forgot *everything* a week later. That type of forgetting is probably permanent, unless the newly learned information was called upon frequently afterward. Declarative memories, therefore, are most impervious to loss when they are used constantly, especially over prolonged periods.

In addition to declarative memories, there is another type of long-term memory that seems practically impervious to forgetting.

Nondeclarative Memory

Nondeclarative memories are stored nonconsciously. They guide our behavior, often without knowledge about where or when an original lesson occurred. They are termed nondeclarative because they cannot be recalled at will, they cannot be "declared." By consensus among researchers, these are the types of memories that *cannot* be brought into consciousness for study, presumably because of how they are stored, or where they are stored, or both.

As an example we could consider little Miss Muffet, sitting there, eating her cereal, when a large spider spins down from a tree. Does Miss Muffet pause a moment to recall every previous spider encounter before deciding what to do? Probably not. Rather, she is likely to be flooded with adrenalin, at which point she summarily takes her leave, leaving the spider hanging.

Fear in response to the spider can be a form of *memory*—that is, a summarized version of previous spider encounters (or stored beliefs about spiders). The strong experience of fear, in association with the spider, a "fear memory" could have remained outside of Miss Muffet's awareness for life, except for the uninvited appearance of the pushy arachnid. Had the spider been kind enough to remain out of sight, the "fear memory" would not have surfaced.

Nondeclarative memories, as we can tell from this example, are *highly* dependent on context. They "pop up" as a means of guiding our behavior. Just like declarative memories, nondeclarative memories are made up of different types, which include procedural memory,[14] skills and habits, conditioning, priming,[15] and possibly, nonconscious attitudes. (Not all researchers compile the same list of nondeclarative memory types. Thus, we will be looking at several that are described frequently and appear to muster some consensus.)

Procedural memory is a type of motor memory, like knowledge about how to ride a bicycle. When we ride bicycles, we can do so without any memory of when we learned. If we know how to ride, we can do so rather effortlessly, even though we may be unable to describe exactly "how" we are doing it, or when we learned. Thus, the "ability to ride" is stored as a "motor" memory (a procedural memory), completely separate from memories about "when we learned to ride." The reason researchers know that the actual procedural memory is stored separately is because a man with amnesia can forget when he learned to ride, where he purchased his first bicycle, yet retain his ability to ride, which is a procedural memory. Conversely, he could have lost his ability to ride as a result of a different type of injury, yet still retain a memory about when he learned to ride.

This example illustrates that memories are dissociated, which means they are stored in separate areas of the brain. (This example also illustrates the networking principle, particularly the aspect of segregation.)

As we saw, there are other types of nondeclarative memories, including "skills and habits," "conditioning," "priming," and possibly nonconscious attitudes. Skills and habits follow the same principles that we just considered, relative to "procedural" memory, and it is not certain that scientists are referring to separate memories. Nevertheless, the same general principles apply to skills, habits, and procedural memories, which we will not discuss further. However, we will discuss "priming," which is another type of nondeclarative memory.

Priming occurs when a behavior is influenced by previous experience, but the priming effect occurs outside our awareness. Priming is probably a combination of several nondeclarative memory phenomena, because it can pertain to either verbal or visual information. To illustrate priming, assume that we are watching a television program that features the ocean on a moonlit night, and during the program, a marketing firm calls to ask what brand of laundry detergent we like best. There is a good chance that we will say "Tide," but make no connection between the televised scene of the ocean and the question about detergent. The word "tide" was primed by pictures of the ocean bathed in moonlight. (By the way, I concede this is a flawed example insofar as we might have said "Tide" only because it is a popular detergent, but "priming" is a well-documented phenomenon.)

Another type of nondeclarative memory is an attitude. Attitudes are the types of memories that serve as the basis of our thinking, but remain largely out of sight. If you were bitten by a dog at the age of three, you might retain a vivid recollection of the incident (which is a declarative memory). However, as you grow older, you would likely encounter other dogs. Eventually, a memory or attitude about "dogs in general" would form. This "dogs-in-general attitude" is an abstraction, and abstractions can form nonconsciously.[16]

Even if you could recall all dog-related encounters over the course of a lifetime, there would still be no memory of the exact moment when a "dog-related" abstraction *first* appeared. That attitude would emerge only in time of need—possibly the first time you saw a large canine while taking an evening walk. In general, when beliefs or attitudes come into existence, their origins are often unavailable for our conscious study. We learn of their existence only through their influence on our behavior.

The sine qua non of a nondeclarative memory is that it influences behavior outside of awareness. In 1993, I was riding a bicycle through the streets of Houston, Texas, attempting to cross a wide thoroughfare. I could see traffic approaching quickly, so when I got to the median strip, rather than stopping to lift my bike up the curb, I tried to jump the curb. My attempt at trick riding was a failure and I was thrown to the ground (fortunately, on the grass covered median). To this day, hard as I try, even with a new mountain bike especially designed for such shenanigans, when I approach a curb *of any height*, my muscles will not respond to my conscious commands. Fear infuses my body and my arms steer the bike away at the last second. I have been largely unsuccessful at overriding my body's memory (fear), even after years of trying.

Now, here's the crux of my story. *I can barely remember the original incident.* I could not begin to tell you which street it occurred on or what part of the city I was in. I have almost no conscious memory of any details, although for several days afterward, my mind was filled with thoughts about the event. The consciously recallable memory of the incident is a declarative memory, stored in a different part of my brain, and it has practically faded from existence. (A conscious memory may only remain because I have repeated this story several times.)

The physical or bodily memory of the incident, in contrast, is as fresh today as on the day when the accident occurred. The bodily memory continues to guide my behavior *in the absence of a detailed conscious memory*, which leads to two essential features of nondeclarative memories.

First, nondeclarative memories often form on the basis of a single episode. As illustrated, a nasty spill on a bicycle can influence riding behavior for life, which is the hypothesized reason for nondeclarative memories. They automatically influence future behavior.

The second essential feature of a nondeclarative memory is that it remains dormant except in time of need, and then surfaces as a strong feeling or motor memory to guide behavior.

Nondeclarative memories do not necessarily fade over time or decrease in strength through disuse. They are the types of memories that humans share with animals because both humans and animals can learn through conditioning, for example, which is a type of nondeclarative memory. Thus, these memories are thought to originate in "older" parts of the brain, from the standpoint of when they evolved. Importantly, then, they have the longest track record for sustaining species survival.

As indicated, nondeclarative memories seem *impervious* to retrieval failure, except in cases of brain damage or brain disease. If you are an accomplished swimmer, even if you have remained out of the water for fifty years, but then accidentally fall into a lake, you may not show the finesse you did fifty years ago, but you will not have forgotten how to swim.

Are Memories Accurate?   Recall the bicycle-riding anecdote from earlier. Your memory of that anecdote is either a short-term memory, stored in working memory, or a declarative memory. In contrast, the event for me is stored in at least two ways: It is a declarative memory and a nondeclarative one. How so? I can recall the anecdote consciously (barely) and entertain my friends with the story. My ability to call forth some of the details is "declarative." Some aspects of the event are stored as bodily sensations, though, as a nondeclarative (procedural) memory.

Even today, as I mentioned, if I am out riding and approach a tall curb, I have a physical reaction that is strong enough to prevent me from trying to jump a curb. Long after my conscious recollection of the details has faded, "older" and "lower" parts of my brain continue to remember the negative aspects of the original experience, with the result that I am unlikely to repeat that mistake again. A so-called "body memory" (a visceral reaction in the pit of my gut or a muscular reaction) stops me from trying to ride up a steep curb. At the moment I might try, my body is infused with fear, so I stop before I get hurt again.

Now, to the crux of this discussion, is my memory of the incident accurate several years later? The answer to this question depends on many different things. When I say "memory," it depends on whether I am talking about the actual details of the event, like where it occurred, what street I was on, the extent of my scrapes and cuts, or perhaps the simple fact that the event occurred at all. Many of the "facts" have largely disappeared from memory.

Since I cannot recall many of the details, is it reasonable to conclude that my memory is no longer "accurate"? When the accident occurred, there were potentially hundreds of memorable details. With time, though, many of those details have been lost. Is a less detailed recollection "accurate"? Clearly, the richness and detail of my memory has faded.

Complex events become stored in component parts.[17] When some details are lost, is "the memory" still accurate? Who can say? It may depend on what "parts" of the memory are lost. With fewer details, the memory clearly loses some of its vividness. If enough details are lost, the memory has been altered; it has *changed*.

Complex life events that involve emotions, thoughts, and behaviors, like a surprise birthday party, are believed to be stored in component parts in different areas of our brains. Whereas you or I can recall a particular history lesson from high school and modify those facts in light of new information, we cannot consciously call forth our emotional reactions to history class and then rearrange those feelings to suit ourselves.

Each of the different types of memories described are thought to be produced and stored in different areas of the brain, and some memories are moved from where they were first processed. People who develop Alzheimer's disease, for example, show greater impairment for recent events during the earlier course of their illness. They may forget about appointments or become lost while taking a walk, forgetting which way they came.[18] Because there is a differential loss of memory, a greater loss for recent events, researchers have concluded that as memories grow older, they are moved to different parts of the brain.

It is true that for many years scientists studied specific brain structures in an attempt to pin down the "seat of memory." They embarked on that path because it was well known that damage to both humans' and animals' brains often resulted in serious memory impairments. Perhaps you have heard about or known of someone who had a stroke and lost some memories as a result. Sometimes, memory losses are extremely specific. Someone could lose their ability to recall certain types of objects, like kitchen utensils, yet retain their ability to name items used for repairing a car. Because of cases like this, neuroscientists hypothesized that memories must be localized in specific areas of the brain.

With improved brain-imaging techniques, like PET scans, and tremendous gains through animal research, it has become increasingly clear that memories can be widely distributed in the brain; in fact, they are now known to exist in all areas of the brain, as we have learned. Some types of memories may be stored in one location; and for other types of memories, different parts of the memory may be distributed in various areas of the brain.

Some remnants of a memory appear to be stored in the respective regions of the brain where the processing occurred during the original event. If you have ever witnessed a minor collision between two cars, the event is quite complex, and so is the memory. During the actual event, you would have experienced sights, sounds, perhaps distinctive odors where the collision occurred; and you may have seen other people standing around. For the brain to put all that together, countless sensory inputs are integrated in various "association" regions of the brain, long before a picture coalesces into "a car collision."

Colors and shapes have to be integrated before you recognize an object as a "car." Since only sensory cells can process sensory information, in all likelihood, it is believed that aspects of the complex memory are stored in simpler, component parts. In other words, part of the complex memory is stored in a visual part of the brain, part of the event is stored in the auditory part, the emotional aspect is stored separately, and so on.

There is no doubt that specific brain structures are required for certain types of memory. Scientists have done a great job in identifying some of the more critical structures. The general area called the limbic system, introduced earlier, often receives the lion's share of credit for many types of memory. Similarly, the contributions to memory from specific nuclei within the limbic system are also becoming clearer. As a general rule, though, it is safest to conclude that different structures of the brain are involved in different types of memory, particularly those of the limbic system. However, memories are stored throughout the brain, and complete knowledge about all areas that contribute to information storage has not been achieved.

The Attribution of Meaning.   There is another critical aspect of memory formation and recall. One of the major principles mentioned earlier incorporated the idea that meaning must be attributed to some types of stimulus before they can be processes for thinking or memory.[19] At any given moment, there are too many sensory inputs for the brain to process. Thus, as a compromise, the brain attends to only those events that it perceives as being important, but importance depends on the *meaning* we give to an event.

Suppose you have never been in the woods, and you see a medium-sized wolf—and twenty years later you still retain a vivid memory of the wolf you saw while hiking in the woods of Colorado. In contrast, a local resident saw the same animal and recognized it as the neighbor's dog and promptly forgot the whole incident. What each person remembers is heavily influenced by each individual's *interpretation*. In a literal sense, what people believe they saw influences what is later remembered.[20]

Many factors contribute to the meaning that an individual attributes to an event, including previous experience, intelligence, familiarity with a situation, and age or level of maturity. Even infants attribute meaning to objects fairly early in life as a precondition of storage.[21] However, the meaning attributed by an infant is vastly different from the meaning given by an adult. Infants and children store information based on their level of cognitive ability, as we have seen, and the very young store information as bodily sensations or emotional reactions, because the parts of the brain that create motor memories and some emotional sensations mature earlier in life than "upper" areas of the brain.[22]

When an immature belief is stored nondeclaratively (outside of awareness), either as a bodily sensation or attitude, that memory can influence behavior for life, completely outside our awareness, so the memory may *never* be updated in the face of increased maturity. As adults, we might avoid cats because of frightening encounters with them early in our lives. We might be so bothered by cats that we decide to seek therapy for our fear. However, in the absence of any memory about the original incident, how would we explain our fear? "I have always hated cats" (which is not true). As we can see, this explanation fails to capture the original event, because the only memory of the event is a "fear of cats."

(No declarative memory of the original biting incident remains.) This is a critical issue because therapists and clients often work together to understand traumatic memories in the absence of specific details. Some therapists interpret the fear, and many clients come to believe these interpretations. Interpretations, though, obviously have varying degrees of accuracy. Thus, when clients and therapists agree on the meaning of a bodily sensation, the interpretation may be close to the mark, or wide of the mark.

When attempting to explain bodily or emotional sensations in words, any verbal description is highly subjective. No two people describe the same internal experience in exactly the same words, and verbal descriptions of emotional events are never exact duplicates of the actual event. Talking about fear *is not* fear (although talking about a frightening situation might trigger fear again).

Ever since Freud first popularized the concept, researchers have been interested in the influence of mood and emotion on our memories. It is now known that a current emotion or mood helps us remember material that was stored at a time when we were experiencing a similar feeling or mood, with some exceptions.[23] This phenomenon is called a "mood-congruent memory." Thus, when we are depressed, we can remember many unpleasant events, ones that we may never even think about when we are happy.

Our level of maturity also influences recall. For example, someone can look back at a father's abusive behavior and, as a mature adult, comprehend mitigating circumstances that could not be understood during childhood. Thus, an adult of forty might look through some old papers and discover that his father had painful migraines for many years, which caused the father to be irritable most of the time. Learning about the migraines years later can cause the adult son to look back on his father differently. As he looks back with added insight, his feelings about his father may change. As his feelings change, however, memories of his father are also changing.

As we can see, our memories are very different from simple snapshots. They involve dynamic processes. We actively reconstruct events, often based on fragmentary information, often based on inferences about what *might* have happened.[24] As humans, we actively construct stories that tie loose pieces of information together, and memories are often brought into alignment with our *current* beliefs about the world. A perpetual optimist will remember past injuries and reinterpret them as "growth experiences." A hostile paranoid will look back on the kindly actions of a passerby who said "hello," and reinterpret that greeting as a failed attempt to beg for money.

When memory functions are viewed from an adaptive perspective, it is not difficult to see how even "distorted" memories could have served important survival functions. If a prehistoric ancestor remembered from his childhood that bears were dangerous, does it matter that the original learning event involved a dog and not a bear? If a distant ancestor believed that he had been threatened by a stranger in childhood, then lifelong "stranger avoidance" could provide an edge of safety when strangers are encountered in the wilderness.

Clearly, our memories do not have to provide us with the accuracy of a photograph to meet our needs. On a day-in, day-out basis, our memories serve us well. However, just as clearly, there is a downside to fallible memories. Ample research has shown how unreliable eyewitness testimony can be, and there have been many cases where innocent people went to jail on the basis of mistaken identity. So, what can we conclude? Our memories have some degree of fallibility, without doubt, but the degree of fallibility, to date, has not been so pronounced as to have jeopardized the survival of our species. This is not to suggest that the fallibility of our memories cannot cause us, or others, extreme harm. If someone has been unfairly convicted on the basis of inaccurate eyewitness testimony, then the fallibility of memory has exacted a high price.

The next major principle, "The Memory Principle," embodies many of the points we have been considering.

Principle 7: The Memory Principle

As humans, we have many types of memories, and different types are stored in various parts of our brains. Further, different types of memories house different types of information. Finally, some memories are more easily forgotten than others, and some memories change over time.

There are few issues more critical to understanding human behavior than memory. Not only do psychologists conduct much of their business on the basis of what people remember about things that have happened in the past, but doctors rely on patients to remember symptoms. Police officers investigate crimes on the basis of what witnesses remember. Prosecutors prosecute on the basis of witness testimony. We give our "word" to others, and the sanctity of our word is based, in part, on what the different parties remember about the original agreement (and this list could continue indefinitely). When it comes to understanding behavior, few factors play a more pivotal role than memory, so I have chosen three very typical examples that illustrate aspects of "The Memory Principle."

IMPLICATIONS FOR UNDERSTANDING BEHAVIOR

Mistaken Meaning

I once attended a banquet held in celebration of several individuals who had successfully "kicked" various drug or alcohol habits. Each celebrant was expected to give a brief testimonial to the gathered audience, which consisted of friends and family members. Some of the celebrants gave what were obviously impromptu comments, others spoke from note cards.

The first speaker likened his addiction to a terrible burden that had caused him tremendous problems in his life. In trying to explain the experience of addiction, he resorted to an unfortunate analogy. He said, "Having an addiction is like being illiterate when you're looking for a job." Shortly after his talk, another man took the podium and talked about how proud he was, and then said, "And I'm proud enough to stand up here, to stand up to those who say I'm stupid because I can't read." (It was a very awkward moment.) The room became hushed and even as a bystander, I felt embarrassed.

The second speaker had felt humiliated by the comments of the first speaker. Afterward, I heard them trying to resolve the issue in the hallway, but they were both fully convinced of the "rightness" of their respective positions. This example illustrates that what we remember is often stored *after* we have made some interpretation. Few of us can recall the verbatim statements of others; rather, most of us form memories conceptually, on the basis of an interpretation. (Although you could argue that this example is really one of a social interaction, the point would be the same.) Memories are formed on the basis of how we each experience and *interpret* a social event, yet try to get two people to agree on what actually happened or what was said.

Let me give another example. Suppose a small child heard her mommy and daddy "screaming" behind closed doors. As you and I know, there might be many reasons for "screams" coming from behind the bedroom door. Children, though, who are developmentally immature, can interpret events only within the context of their limited life experiences. Since, hopefully, children have not had firsthand knowledge of adult sexual activity, they can only interpret the "screams" from their limited repertoire of world experiences. That limited basis of experience, though, is also the basis of the memories that are formed. Thus, the child at the bedroom door may forever be convinced that "daddy was hurting mommy," or vice versa. This experience could coalesce into a nondeclarative memory, and for years to come, the child might believe that "daddy was being mean to mommy," and the memory could even affect the child's life, if it were traumatic enough.

When people attempt to describe unpleasant or unwanted feelings that arise from past events, what interpretation might be placed on vague or emotional reactions, in the absence of any memory about where those feelings came from? As we saw, clients and therapists often attempt to make sense of past encounters through what are essentially educated guesses—hypotheses. Hypothetical explanations about the past, however, vary in accuracy, and few can ever be proven.

Traumatic Memories and Repression

No discussion of memory would be complete without some consideration of traumatic memories and repression. You may recall the example from earlier about two therapists who had been sued in California. According to the newspaper, they had "convinced" someone that an occurrence of sexual abuse had occurred, but an opposing attorney argued that the therapists had "planted" those memories. After a guilty verdict, one therapist was quoted as saying, "Repressed memories are a reality."

Among professional psychotherapists, including psychologists, social workers, counselors, and psychiatrists, there is a very large subset who embrace strongly the concept of "repression." Repression is most often associated with Sigmund Freud, whose theory refers to a well-documented memory phenomenon. That is, people often forget the details of a traumatic or sexual event. Observers often wonder, as I have done myself, if someone witnessed a murder at the age of seven, or was sexually abused throughout childhood, how could the person forget such a thing? Nevertheless, within the psychiatric literature, there are literally thousands of examples of people who appear to have lost their memories of highly upsetting events. Although some "experts" have dismissed these assertions out of hand, the phenomenon is too widespread, and too many therapists have seen evidence of "lost memories" to simply label the evidence as unsubstantiated.

Repression, however, refers to a very specific theory about why some memories cannot be recalled. Repression was elaborated by Sigmund Freud to explain someone's apparent ability to forget an event that had so much impact. He believed that repression required a constant expenditure of energy to keep the objectional material out of awareness as a protection against overwhelming anxiety.[25]

Even though Freud was a physician, at the time of his writing the brain was largely a mystery in many important respects. Thus, he relied on figurative terminology to explain his theories, and one hallmark of his theory involved the notion of the "unconscious."[26] Although the term "unconscious" is used by many psychotherapists, it does not denote a specific part of the brain; rather, like repression, the unconscious is a figurative term referring to a function. According to Freud, any material that was highly objectionable—material that was likely to overwhelm an individual with anxiety—was repressed; that is, it was kept out of awareness in the unconscious. (Freud was constantly revising his theory, so you have been provided with a piece of his earlier theory, and only the nickel version at that.) Many books have been devoted to the topics of the unconscious and repression. Also, to reiterate, we should not equate the unconscious with nonconscious processes, as I have sloppily done on many occasions, because the terms denote different phenomena to some professionals who are sticklers for detail.

Now, at the risk of being branded a heretic, or worse, and thrown out of my profession, I would like to take issue with the concept of repression, based solely on the material and principles already discussed in this chapter. In my way of thinking, the concept of "repression" does not square well with emerging research on memory.

As sure as I am sitting here, I am convinced that people are unable to consciously remember the details of some traumatic events. I am not quibbling with that assertion. As a social scientist, I believe the loss of details for traumatic events has been well documented. However, I believe that there are better ways to explain the loss than resorting to the concept of repression. In my judgment, the concept of forgotten traumatic memories can be explained easily by emerging brain research on memory without relying on the concept of repression. One of the most common examples of a repressed memory involves childhood sexual abuse. As adults, "abused" individuals either cannot remember the occurrence or cannot remember important details of the occurrence, or they recall the event years after it happened, wondering how they could have kept it out of their minds for so long. You may recall the case of a former Miss America who went on public talk shows and openly discussed allegations of sexual abuse toward her father after he had died. The fact that these types of memories are often remembered after the perpetrator is no longer around is not uncommon in my experience. So how can these cases be explained by the normal memory principles we have been discussing?

First, traumatic memories are *not* lost, per se. In the bicycle example provided earlier, we saw that the essential feature of a trauma can guide behavior for life, even after the details of the event have been lost. The effects of a trauma only appear lost because we cannot consciously recall the details of the event. Why is that? Complex emotional events, and perhaps other memories as well, are broken up and stored in component parts, according to the researchers. The conscious details are declarative memories, and declarative memories are often lost, unless we recall those details on a regular basis. However, many of us choose to divert our attention from unpleasant details, so it is not surprising that people often forget the declarative aspects of a traumatic event, especially events that occurred long ago.

We also learned that long-term declarative memories cannot be recalled without the right "key." Whenever we access a declarative memory, we do something that unlocks the memory, figuratively speaking, but no one knows exactly what that "something" is. We do know, though, that the brain is highly attuned to context, and memories are released in situations that are similar to the original event. For highly unusual, age-inappropriate experience, how would a young child's brain interpret the act of sexual abuse and then store it? How would a young child's brain encode that experience? For conscious memories to form, we have to attribute meaning to the experience, but what meaning might an immature girl attribute to an event that is so far outside of a normal childhood experience? We know little about how the brain stores information to begin with (in comparison to *all* that is not known), much less how it stores highly unusual events. (By definition, a traumatic event is so out of the ordinary that one would not normally come across environmental cues that would be reminiscent of the event.)

Also, as in the case of Miss America and her father, when older men abuse young girls, especially adult relatives, they invariably make up a cover story, like "I do this because I love you," or "this is our special secret," and so forth. Because of the child's age, it may not be possible for her to critically analyze these messages, especially when the message is given by an adult who is trusted. As we learned, memories may be highly subject to reinterpretation, especially when contradictory information is provided by an individual who is trusted. Thus, if Daddy says "this is an act of love," the child may be unable to question that assertion. Also, the brain, especially a young brain, is extraordinarily malleable, so it can be persuaded to believe anything.

The concept of repression is not tenable in light of new memory research. First, it does not appear as if the brain has excess energy. The brain cannot store its own glucose or oxygen, so it is hard to imagine that a mechanism evolved that, by definition, is energy consumptive, unless that function provided an *enormous* advantage to an individual. (It is true that thinking processes require a large supply of energy, but there are tremendous payoffs for original thought processes.) It is hard to imagine a reason why the brain would evolve an energy-consumptive system, like repression, for the purpose of ignoring its own output.

Second, we humans can think about the most objectionable material anytime we want. All of us can recall many events in our lives that were extremely painful. Thus, the notion that we need to be protected from "objectionable material" is hard to justify. Viewed from the perspective of our distant past, it is hard to imagine that our prehistoric ancestors had the time to sit around and worry about their inner experiences and childhood traumas. (Repression was popularized by Freud, in all likelihood, to reflect the times and beliefs of his day, but it is more difficult to see the need during prehistory when individuals were in the constant company of companions and probably too busy with basic survival demands to sit around and be paralyzed by anxiety.)

Finally, the most convincing argument against repression is that many, perhaps most, of our memories cannot be recalled, as we have already discussed. Since many nontraumatic memories cannot be recalled consciously, it seems *unlikely* that the brain would have evolved a separate set of rules for traumatic events. It is more likely that the brain has one set of rules for each type of "memory system," and within that system, it does not matter what type of information is stored. This view is certainly the most consistent with respect to emerging brain research, although I concede it is at odds with traditional views of repression.

In sum, I believe that various *normal* memory principles can be used to explain why traumatic events are so difficult to recall. First, aspects of the event are stored nondeclaratively as feelings or motor memories, not subject to conscious recall. In turn, those feelings may be difficult to describe, and any verbal description may never fully capture the original experience. Second, the declarative memory component will fade with time, unless the memory is called upon frequently, which is unlikely under the circumstances. Traumatic events are the types of events that people do not wish to recall. Further, declarative memories appear to require some mnemonic device for retrieving them, but when a memory is formed by highly unusual circumstances, particularly when the victim is a young child, the mnemonic key may elude the individual for years. Finally, most of the brain's volume is devoted to nonconscious processes, so we do not need to rely on a "theory of repression" to explain why some memories remain out of sight.

Age and Forgetting

The aging process is a topic that is near and dear to most of us. There appears to be a common perception that old necessary means memory loss. For many types of memories, recall depends more on use than age. Thus, people who practice using their intellect may see few declines in their memory as a result of age. The adage that applies to physical fitness also applies to memory, "use it or lose it." This does not apply to an actual brain disease, like Alzheimer's disease, but relatively few people will develop Alzheimer's.[27] For most of us, therefore, memory processes can be kept vibrant throughout life.

Further, as we have seen, nondeclarative memories may persist for a lifetime, even if we do not use them, as exemplified earlier by the person who fell in a lake after a fifty-year absence from swimming yet still retained the ability to swim.

CONCLUSION

Among researchers, there is an increasing consensus that we have many different types of memory. However, researchers do not agree about the most desirable ways to describe or label those different memory processes. Biologists would prefer to simply describe memories in terms of brain structures and neural pathways, but for many memory phenomenon, we are years away from knowing where those pathways might reside within the brain. Thus, general figurative terms are used; those terms often describe the ways in which memories manifest themselves, telling us little about underlying brain processes.

Speaking figuratively, there are many memory systems, including: (1) working or short-term memory, (2) intermediate-term memory, and (3) two general types of long-term memory, declarative and nondeclarative. Each category of memory is different from every other category in terms of the type of information stored, the causes of forgetting, the possibility of conscious recall, and the possibility that the memory's meaning will change over time. The major attributes of memory are summarized in "The Memory Principle," reiterated here.

PRINCIPLE 7: THE MEMORY PRINCIPLE

As humans, we have many types of memories, and different types are stored in various parts of our brains. Further, different types of memories house different types of information. Finally, some memories are more easily forgotten than others, and some memories change over time.

* * *

As if this chapter did not contain enough twists in the road, let me add one more. The distinction between memory and learning is not clear-cut. Whatever has been "learned" has also been stored in memory. The reason we use two different terms is because both terms, memory and learning, are used in the scientific literature, and each refers to different aspects of the same general phenomenon. So, as we embark on an investigation of learning, we will discover facts that also pertain to memory.

CHAPTER 7

THE WELL
CONDITIONED BODY

Imagine you are on a camping trip in a national forest. You have been hiking all day, it is early afternoon, you are tired and hungry. Arriving back at the campsite, you start making dinner. Unfortunately, you are low on water, so you start walking toward the nearest faucet, about a quarter-mile away. On your way, you notice a shortcut and do *not* even consider *not* taking it. After all, you are beat. No sooner do you step on the path than you notice a very large rattlesnake catching some late afternoon rays. Most of us would waste little time in turning around, and some of us would avoid taking shortcuts from then on (or even give up camping altogether).

Looking beyond the inconvenience and fear, think of this situation as a lesson. In one brief instant we have learned to avoid a potentially dangerous shortcut forever. Now, compare this form of learning with an everyday occurrence, like memorizing the name of a stranger to whom we have just been introduced. In the case of the name, we may or may not remember, but who could forget a shortcut that featured a rattlesnake?

We could argue that learning to avoid a path with a snake is not all that useful, especially if we live in a city. Moreover, if we decide to stay in that campground, what is the likelihood that the snake will be in the same place tomorrow? If there was a brush fire later that night, the path could come in very handy, so the fear of taking it might be a handicap.

Nevertheless, as a type of learning, a brief exposure to the snake may have resulted in a lesson that will never be lost, so for pure efficiency, this kind of learning cannot be beat.

Here is another common example of learning. Jack was a 34-year-old salesman who had a lifelong problem controlling his weight. He loved all sorts of fattening foods. After a recent job promotion, his weight really ballooned. The problem was that, with that much more money, there were no financial constraints on how often he could go to his favorite donut shop. So, whenever the urge struck him, he jumped into his car and off he went. The habit of jumping into his car whenever he wanted a donut is a learned behavior, one "reinforced" by the donuts at the end of the line.

Both of the foregoing examples seem quite different, but they have one thing in common: They are both types of learning that belong to a broad class of learning called "conditioning," which we will discuss in this chapter. However, before getting to the specifics, I should stress that the differences between memory and conditioned learning are not clear-cut. Thus, any broad conclusions that apply to conditioning may also apply to memory.

Because learning is a brain process that cannot be observed directly, it is generally inferred, most often through a change in behavior.[1] However, learning is not the only source of a new behavior. Behavior can change as we grow older, and these changes result from maturation (development), which we have discussed previously.

Learning *always* involves a change in the brain. For some forms of learning, if not all, it is thought that information is stored through changes in synaptic connections, but the actual location of learned material within the brain depends on the material.[2] There are storage sites throughout the brain. However, scientists have not discovered all the places or ways in which material is stored.

Ironically, many learning theories have intentionally omitted any reference to the brain, even though the brain is the undisputed location of human learning processes. This may seem odd to us, so a brief look at a piece of psychology's history will help to clarify the matter.

Psychologists have long been interested in behavior, but at one time, issues of the "mind" were thought by some to be impediments to understanding behavior.[3] In the early 1900s, the technology did not exist for the types of brain studies that are possible today, so in an attempt to put psychology on a firm scientific foundation, to make it "objective," some psychologists argued that concepts like "consciousness" had to be abandoned in favor of a focus on *observable behavior*. This is not to say that psychologists were not interested in learning, but they studied only what they could directly observe, like what happened immediately before, during, and after a specific behavior. If something could not be seen and measured directly, it had no place in a "science of psychology." Thus, many theories of learning and behavior proliferated without any mention of the brain.

The field of psychology has been subjected to some blistering attacks,[4] partly because psychological theories of learning omitted a discussion of the brain's role in behavioral phenomena. However, in defense of experimental psychology, it can be argued that it has nothing for which to apologize. Experimental psychology has uncovered principles of learning that have had enormous influences on education and every other field that is interested in behavior. The influence of behavioral psychology is widespread in our culture. Many of the principles uncovered by experimental psychologists have proven to be extremely effective methods of modifying a wide variety of problems and those same principles have been very useful in all types of teaching endeavors. Behavioral theories are well imbedded within diverse educational settings, and experimental psychology has provided us with terms and concepts that permeate our language and culture, like "conditioning," "reinforcement," "behavior modification" (B-Mod). In fact, what I discuss here leans heavily on the contributions of experimental psychology because the findings from that branch of science have provided us with enormous insights into behavior.

CONDITIONING

As a general type of learning, conditioning encompasses several varieties, just as memory consists of different types. It is difficult to make generalizations about conditioning because each type follows slightly different principles. However, a few generalizations can be made.

First, conditioning unfolds along predictable principles. When those principles are well understood and applied, a "conditioned" response can be created, often whether we want to be "conditioned" or not. Some forms of conditioning can occur even if we try to resist, and some forms can occur outside our awareness.

Some of us may have attended elementary schools years ago and may still have memories of the experience. Our emotional reactions to school, whatever they may have been, occurred pretty much on their own. Even though we were aware of the conditions within our classroom, we were unable to keep from developing an overall emotional response to school. Eventually, we may have developed an emotional reaction to school that we experienced as a feeling. This example illustrates "emotional conditioning," which is not dependent on our conscious involvement or our awareness that we are being affected in any way.[5]

When it comes to conditioned learning and consciousness, the two phenomena have an unusual relationship to each other. When the neocortex evolved, we acquired awareness or consciousness. Awareness, in turn, provided us with the ability to know about *some* of the brain's other functions, but not all of them. Thus, we *can* be aware that we are being conditioned, in some instances, but our awareness may not prevent the learning from taking root. However, awareness is not necessary for conditioning to occur. Although this may seem confusing, as we get into specific types of conditioning, this point will become clearer.

In comparison with conscious learning processes, conditioning is very old. Assertions like this are made because conditioned learning among humans follows many of the same rules and principles that are observed in animals, and because both humans and animals share the same brain structures that are required for conditioned learning.[6]

Since we share conditioning with many or most animals, there are several implications to this general rule, the first of which we have already learned: Human conditioning follows the same rules, patterns, and neural networks as those found in "lower" animals. Shock a mouse a millisecond after it begins a new behavior, and you will change its behavior for life (or kill the mouse)—and the same holds true for humans. A good jolt of electricity delivered milliseconds after we begin to eat a juicy steak may influence our dining behavior for life (although I do not recommend electricity as a diet aid).

Second, for all animals, humans and nonhumans alike, conditioned learning imparts some of the most critical survival lessons, often early in life. For humans, conditioned learning helps to teach a child about the dangers of the world before "higher" cognitive processes have developed. If a toddler begins to wander into the street and his mother shrieks "NO!" in her loudest voice, her yell will be experienced unpleasantly by the toddler, and hopefully he or she will not repeat that mistake again.

Because there are different types of conditioning, and because one general type is quite different from the other, it is probably best to just jump in and get down to specifics.

Classical Conditioning

"Classical conditioning" is also called Pavlovian conditioning because it was discovered accidentally by Ivan Pavlov, a Russian physiologist.[7] Classical conditioning is described by a bewildering array of terms

that can be confusing, like "conditioned stimulus," "unconditioned stimulus," "conditioned response," and "unconditioned response." It is much
easier to understand classical conditioning, however, simply by looking at
what Pavlov did.

Early in the twentieth century, Ivan Pavlov was studying digestion,
so he developed surgical techniques that allowed him to study more easily
the salivation of dogs. He performed a surgical procedure on the dogs,
which diverted their saliva through a tube to the outside of their mouths,
where the exact amount could be accurately measured.[8]

Pavlov observed that dogs salivated when they were presented with
food, so he considered their salivation to be a natural response (an "unconditioned response"). Pavlov's big discovery, however, came when he observed that his dogs also salivated at times when there was no apparent
reason. Specifically, he noticed that dogs might salivate when they hear the
approach of their caretaker, that is, the footsteps of the kennel attendant
could cause the dogs to salivate. With his curiosity aroused, Pavlov studied the phenomenon more closely. He soon learned that the dogs could be
made to salivate in the absence of food, if a neutral "stimulus," like a
sound or light, was "paired" with food. That is, if a light is turned on every
time a dog is presented with a bowl of food, eventually the light itself
will cause salivation (even when the food is withheld).

Since the time of Pavlov, researchers have discovered many things
about classical conditioning. First, most animals, including humans, can be
classically conditioned. Even people with amnesia often retain classically
conditioned responses, often after losing other types of learned information.[9] (This also illustrates that different memories or types of learning are
stored in different areas of the brain which further attests to the networking principle.)

Classical conditioning can occur very quickly, although it is popularly
characterized as requiring several trials before it takes root. In truth, it can
occur after only one or two pairings between two stimuli.[10]

Apart from both psychological and biological interests in the phenomenon, classical conditioning has wide ramifications for human behavior.
"This fundamental form of learning is critical for adaptive behavior. Many
organisms, including humans, depend upon relatively innocuous signals
to warn them of impending significant events. Survival may well depend
on identifying those signals accurately."[11]

Even though some classically conditioned responses are stored within
"older" regions of the brain, like the brainstem and cerebellum, classical

conditioning plays a big part in influencing behavior, as we have just seen. Thus, knowledge about this learning system can provide both psycho-therapists and educators with powerful techniques for changing some behaviors and treating certain problems. To understand how, we should examine a famous experiment in the field of psychology, which describes an experiment that would be unethical today (if not illegal).

During the early 1900s, a well-known American psychologist, John Watson, was studying phobic behaviors in humans. Phobias are intense fears that are often considered "irrational." An intense fear of public speaking, for example, might discourage us from taking a well-deserved job promotion. Thus, phobias can pose significant problems for some people. Watson believed that many "abnormal" behaviors could be ex-plained through classical or Pavlovian conditioning, which is why he wanted to study the matter.[12]

To test his theory, he *created* a phobia in a small child named Albert. Watson presented Albert with a white rat, and just as Albert reached to touch the furry little creature, a loud, unpleasant noise was made. After only a few "pairings" of the noise and the rat, Albert learned to be terrified of not only rats, but of other white furry items as well, like rabbits and even a white fur coat. Unfortunately (or fortunately), depending on your point of view, Albert's mother withdrew Albert from Watson's reach before Albert's phobia could be eliminated.

Years after Albert's experience at the hands of Watson, one of Wat-son's students, Mary Carver Jones, discovered a technique for successfully eliminating conditioned fears.[13] Like her teacher before her, she first cre-ated a phobia in a child, but afterward, she eliminated the phobia. While a child (with a conditioned fear) was eating, and presumably feeling relaxed or content, Jones brought a rabbit into the room where the child was dining. At first, Jones remained at a distance from the child. Very gradually, she brought the rabbit closer and closer. Eventually the child lost his fear of the rabbit.[14] The child, little by little, learned to feel no anxiety in the rabbit's presence.

Another example of classical conditioning was provided earlier in this chapter. In the snake example, when we first started down the shortcut, the path did not produce any particular response. Once a poisonous snake was observed on the path, though, a fear of snakes became associated with the path, at which point the path itself came to be the cause of fear (even in the absence of a snake).

As we have seen, classical conditioning serves vital functions for both humans and animals alike. We should not feel that classical conditioning is always a nuisance. We have learned to associate a red light with possible danger, and that association might save our lives some day. (Admittedly, we also have a conscious knowledge about the meaning of red lights, but we have a classically conditioned association as well.)

Generalization. One important aspect of conditioning is "generalization." This occurs when a single "stimulus," like the path in the rattlesnake example, "generalizes" to other places. We might return to the city after the camping trip, and thereafter avoid *all* paths. Our fear of one path in a mountain campground is said to have "generalized" to all paths. This concept is important because we can generalize from one situation to another and may do so without awareness.

We might have had a particularly bad experience at the dentist's office when we were younger. Perhaps the dentist began drilling on a tooth before the anesthetic had completely taken hold. The pain of the experience could have caused us to fear the dentist, who happened to be wearing a white lab coat. Long after forgetting about the specific office visit, our "learned" fear could then generalize to all people wearing white lab coats, thus preventing us from seeing other types of doctors.

Classical conditioning involves a type of learning by which two events are learned to be associated with each other, so that a "neutral" event (stimulus) comes to cause a behavior that it had not caused before the association was made. The white lab coat in the earlier example became associated with fear and pain.

Brain Studies. At the biological and neurological levels, more may be known about various conditioning systems than any other type of learning. With some types of classical conditioning, in some animals, specific neural pathways have been identified and completely mapped.[15] In fact, progress made in identifying specific, classical conditioned pathways now serves as a model for what biologists would like to achieve for all learning and memory systems.

It is now believed that conditioned fear is dependent on a brain structure called the *amygdala*, which is typically described as part of the limbic system.[16] (Although the amygdala is referred to as if it were a single structure, it is actually composed of several nuclei.) To understand the importance of conditioning and the amygdala, we can look at a common behavior among monkeys. Even though normal monkeys appear "pleased" at the sight and smell of food, like bananas, a love of bananas is learned through conditioning, and the amygdala is believed to be the brain structure responsible. It helps monkeys, and presumably us, to learn to associate a stimulus (like a banana) with the banana's reward value (good taste and a full stomach). Numerous experiments have shown that without the amygdala, the monkey will not learn to associate the banana with its reward value.[17]

The cerebellum, hippocampus, and amygdala are all involved in specific types of classical conditioning, but each contributes a different function, that is, each may underscore a different type of conditioned learning.[18] It is beyond the scope of this book to describe the roles of these brain structures in greater detail, beyond a general statement about where they reside. The cerebellum is in the lowest back area of the brain, in the hindbrain. The hippocampus and amygdala are higher in the brain, in the limbic system.

Modifying Conditioned Learning.   We have seen that many researchers consider memory and learning to be essentially synonymous, insofar as they both refer to storage mechanisms within the brain. Earlier, we considered the issue of whether memories were reliable, whether they could change. We concluded that they *do change* and discussed some reasons why. Now, we will look at the same phenomenon, changing stored information, with a different twist—*with the intent to change the data on purpose.*

People rarely ask if they can change a memory, presumably because memories have been described like photographs with the assumption that they are unalterable sketches of past events. (It is true, of course, that some people wish to forget painful memories, but in my experience, they do not ask to have those memories altered.) In contrast to memories, people often wish to change certain types of learning. We often hear of people who wish to change a "bad habit," and habits are observable behaviors that arise from a type of conditioned learning. In any case, the principles presented here for changing conditioned behavior could also be applied to changing many nondeclarative and "emotional" memories.

A great deal of research has focused on how to alter conditioned patterns—far too much for me to review in the space we have, so I will describe some general procedures and point out what appears to be the common denominator contributing to change in each case. Many of the following techniques are used by psychologists or psychotherapists in order to modify a whole range of unwanted fears and behaviors. Although it is probably safe to conjecture that most of these techniques are confined largely to research, treatment, or educational settings, any of us could read about these principles and safely apply most of them to our own lives.

Almost twenty years ago, when I was younger and more foolish, I had the unfortunate habit of smoking cigarettes. At the time, I was taking an undergraduate psychology class in "learning theory," which focused on various principles of conditioning. Completely on my own, I adapted those principles and stopped smoking. What I am suggesting is that many conditioning or "counterconditioning" techniques can be undertaken without professional help, but I should caution you, there are some procedures that *definitely* require profession guidance, so before embarking on any change program, you should at least consult with a pro to get a second opinion.

Earlier we learned about Mary Carver Jones and her success in eliminating a conditioned fear by gradually bringing a rabbit a little closer to a child, until he could pet it with no fear. Jones's technique is valid today to eliminate a phobia.

There are those who wish to learn to swim, but are unable to do so because of an overpowering dread of water. There are, however, swimming programs that teach these individuals to swim, first by eliminating any fear of water by employing techniques almost identical to those used by Jones. A student gradually moves closer to the water, over a period of time, always remaining relaxed and comfortable. Eventually, the student can sit in shallow water and remain comfortable. Once a high degree of relaxation has been achieved while sitting in shallow water, the student then moves into *slightly* deeper water. Eventually the fear of water subsides, at which point the swimming lessons can commence.

The work of Mary Carver Jones served, in part, as the basis of a very powerful therapeutic technique called "systematic desensitization," which is attributed to Joseph Wolpe, who patterned his work on the findings of Jones.[19] Systematic desensitization is a two-staged treatment process. First, a client is taught how to relax deeply. (There are many relaxation techniques, and most book stores or libraries have books or tapes that teach different methods of relaxation.) It takes only a little practice every day, over the course of a few days or weeks, to learn to relax deeply. Once an individual has learned to relax fully, the second phase of treatment begins. For the sake of illustration, consider the case of someone who is afraid of flying. Prior to, or in conjunction with, the relaxation phase, the therapist interviews the client to learn as much as possible about the situations that cause the fear (in this case, traveling in a conveyance that looks too heavy to get off the ground). Next, over the course of many sessions, the therapist verbally describes a part of the fear-producing situation to the client, but all the while, the client remains very relaxed.

To begin, the therapist describes a small vignette that relates indirectly to flying. On the first day of treatment, the therapist might say, "As you sit there in a deep state of relaxation, you are also thinking about taking a business trip." Gradually, over time, the vignette becomes more detailed, up to and including a description of flying. All the while, the individual learns to imagine the trip, while remaining in a state of relaxation. (I have used systematic desensitization with many clients, and it is an extremely effective means of eliminating some fears.)

Although conditioned fears can be eliminated through verbal presentations and imagery, as described, researchers have devised even more effective techniques.[20] The real crux of eliminating a phobia involves a principle that has been known for many years (and even embodied in the wisdom of many cultures): "Face your fear." Whatever we can do to bring ourselves face-to-face with our fears (as long as the frightening situation is not truly life threatening), and the longer we can remain in the situation, the greater is the likelihood that our fear will dissipate. This works as a general rule, though I have omitted many of the specifics here. The technique of facing our fears without escaping is called "in vivo flooding," and it has proven itself to be a successful technique for treating many types of learned fears.[21] Sooner or later, if a conditioned fear is to be eliminated, it must be faced. (In vivo flooding is a technique that should not be attempted without help. A trained professional should be involved in this approach, and of course the individual must want to be cured and consent to this form of treatment.)

There are other types of conditioning, which we will turn to shortly, but in the meantime, we can consider the first of two "conditioning" principles presented in this chapter.

PRINCIPLE 8: THE CLASSICAL CONDITIONING (LEARNING) PRINCIPLE

Classical conditioning is a type of learning that can occur outside our awareness, and one particular form of classical conditioning supports "emotional

learning." As a type of learning, classical conditioning cannot be altered through
reasoning, but it can be changed through specialized "counterconditioning" pro-
cedures.

To sum up, classical conditioning is a form of learning in which humans or animals "learn" to associate two events together. Conditioning, in turn, is an "older" form of learning, in that we share it with many animals. Because of the brain structures that underlie conditioning (some of which have been identified), classical conditioning does not require our conscious knowledge for it to occur. Nevertheless, it is an important type of learning for humans and animals alike.

Operant and Instrumental Conditioning

Operant and instrumental conditioning contribute to a wide variety of behaviors. Although you may be unfamiliar with the technical jargon concerning these forms of learning, you probably know a great deal about these phenomena already. Nevertheless, some review of general concepts and terminology will probably be useful, so we will begin our discussion with a common example.

If I put $1 into a soda machine and receive a cool can of my favorite orange drink, I am far more likely to do the same thing at a later time than if I had lost my money and received nothing. By receiving a cold drink after depositing my money, my behavior has been "reinforced." Whether we receive sodas after depositing our money, or earn extra dessert for finishing our vegetables, our behaviors have been "reinforced" by the beverage or dessert, respectively.

Both operant and instrumental conditioning are dependent on reinforcement, but these two forms of conditioning are slightly different from each other. If there is no limit on the amount of reward I can earn as a consequence of my behavior, then we are talking about *operant conditioning*. For all practical purposes, I can get as many cans of soda from the machine as I want. *Instrumental conditioning*, in contrast, occurs when there is a limit to how much reinforcement can be earned. For example, when a child is told that she can have ice cream for dessert if she finishes her broccoli, there is a limit to how much ice cream she can earn.[22]

Both forms of conditioning, as we can see, rest upon a well-established principle: Behavior will *increase* or *decrease* by what happens immediately after the behavior occurs. Because of the close similarities between operant and instrumental conditioning, we will only concern ourself here with operant conditioning, but be aware that the principles we are discussing generally apply to both forms of conditioning.

Behavioral psychologists draw a distinction between two types of reinforcers, which is important because we can use knowledge about reinforcers to change behavior, whether it is our own or that of someone else. "Primary reinforcers" are those things that have a tendency to reinforce our behavior, like food or water. "Secondary" reinforcers also have the ability to reinforce behavior, but only because they have become associated with primary reinforcers. The most common example of a secondary reinforcer is money. None of us work for money for its own sake; rather, we do so because of what it can buy for us. We can use money to buy food or other basic necessities. (Incidentally, secondary reinforcers receive their value through the process of classical conditioning.)

Primary reinforcers are usually tied to some biological or social need. They are natural reinforcers, like food, beverage, relief from pain, and possibly companionship, affiliation, and recognition. (Some experts would argue that companionship and recognition are not primary reinforcers, but I would disagree with them, and we will see why later.)

At the risk of getting bogged down in jargon, some key terms *should* be explored, partly because they are used in everyday language, and partly because some terms are almost *always* used incorrectly, by laymen and many professionals alike.

Earlier, I stressed the term "reinforced" to draw our attention to the word, because it is the preferred term of the experts. However, most of us would probably use a more common term, that is, "reward." Reward is a no-no according to scientists because what might be rewarding to one person can be disgusting to another.[23] To get around this sticky problem behavioral scientists use the term "reinforcer," and they have given it a precise meaning: a *reinforcer* is anything that *increases* the likelihood that a behavior will occur again. We stick our coins into vending machines with the hope that we will get a cold can of soda on a sweltering day. If we get what we paid for, the cold soda is a "reinforcer."

Another issue of terminology, one that is almost universally misunderstood, is the concept of negative reinforcement. By definition, *negative reinforcers* also *increase* the likelihood that a behavior will be repeated, but in a different way than a positive reinforcer. How could this be? Negative reinforcers *increase* the likelihood that a behavior will recur when they *remove* something unpleasant. If I take an aspirin for a headache, and if the headache disappears as a result, the aspirin was *negatively reinforcing* because it took my headache away. Thus, I am far more likely to repeat the behavior of taking an aspirin at the first sign of a headache.

One reason why people drink alcohol is to lessen feelings of anxiety in social situations. By taking a drink, anxiety is quickly eliminated, so the "act of drinking" is *negatively reinforced* (because drinking eliminates the anxiety). Thus, drinking behavior is likely to occur again.

Many people, including many professionals, commonly use the terms "punishment" and "negative reinforcement" synonymously, yet they are *exact opposites*. As we have learned, negative reinforcement *increases* the likelihood of a behavior. Punishment, in contrast, by definition, is anything that decreases the chances that a future behavior will occur.

There are two types of punishment—positive and negative.[24] Positive punishment occurs when it is applied in order to *decrease* a behavior. A swat on the rear end, seconds after a child runs into the street, is the quintessential example of punishment. Negative punishment, in contrast, involves the act of removing something to *decrease* an unwanted behavior. We take television privileges away from children who do poorly in school.

Many years have been devoted to research on punishment and its effectiveness on behavior. Contrary to what many people appear to believe, *punishment does not teach new behaviors* (with one exception). If a child is suspended from school for misbehaving, suspension per se, is a form of punishment. However, suspension can only be effective (in and of itself) *if* the child knows how to behave properly and is only withholding the proper behavior. Suspension, as a punishment, cannot teach a new behavior if the behavior has not already been mastered by the child. In other words, new behaviors must be learned, so they must be taught. Of course, if the behavior we are trying to teach is the "withholding of a bad behavior," then punishment *might* be effective. Quite often, though, punishment is used when our goal is to make someone behave in a *different* way, but with both children and adults, it is often the case that the individual does not know an alternative behavior that will accomplish the same need as the "misbehavior." That said, we can examine the conditions that *increase* the effectiveness of punishment for eliminating or suppressing unwanted behaviors.

Generally, three factors determine whether punishment is effective.[25] First, punishment is most effective when it is "severe." (In a civilized society, though, we may have problems with this concept, but this is what the research suggests.) Second, punishment, to be effective, must be administered consistently. One reason strict laws may not influence behavior is because people may rarely get caught for their misdeeds, in comparison to how many times they "misbehave." We might run a red light thirty times and get caught only once. Punishment for running a red light would be more effective if we were caught and fined every time. Finally, punishment is most effective when it occurs close to the time when a misdeed occurred.

In general, when it comes to "conditioned" learning, an association or linkage must be made between a behavior and an outcome. The association does not have to be consciously understood, but one of the more important factors for ensuring that linkage occurs is through timing or contiguity.[26] *Contiguity* is the degree of association between two events, like behavior and outcome, relative to *time*. It has often been thought that timing is everything in conditioned learning, and it *can be* very important, but the issue of timing is one of those scientific questions that continues to undergo extensive research, without a clearly established law having been discovered. Generally, it appears as if operant conditioning works best when reinforcement or punishment begin closely on the heels of a behavior.

Because the concept of reinforcement is so important for maintaining, teaching, or changing behavior, it is worthy of another look. Reinforcement is so much broader than just food or drink. Also, we can even reinforce ourselves, not to mention others, so the knowledge that we are reinforcing behavior does not necessarily weaken the effectiveness of reinforcement (or punishment).

If we are trying to change our own behavior by using reinforcement, it helps to have an extensive knowledge of what is reinforcing for us. During this book project, while under a time crunch, one of the ways I could "make" myself work harder was to allow myself a ten-minute reinforcement for every one and a half hours of work. (Having a good book to read worked consistently. Reading, in this instance, was *negatively reinforcing* because it removed tension.)

Since we are social creatures, many of us will work long and hard to be around others. We will also work hard to achieve goals that help others or help family members. People and animals will also work long and hard to *stop* something that is unpleasant, whether it is walking three miles in the rain to a drug store for a bottle of aspirin to stop a throbbing headache, or working long and hard to improve the safety of one's neighborhood.

When changing behavior, a key concept is "extinction." Extinction occurs when reinforcers are consistently withheld. If we engage in a behavior long enough and never receive reinforcement for it, sooner or later we will stop. If we were to go to work for weeks on end and never get paid, sooner or later we would stop.

If we are trying to change a behavior, or that of someone else, there are two more concepts involved with conditioning that are useful. Few people show the "perfect" response when trying out a new behavior. When we first entered school, for example, we probably did not perform at the same level as older, gifted students. Should the school have withheld A's until our senior year? If it had, few of us would have finished school. Essentially, the school "shaped" our behavior. "Shaping" involves the process of reinforcing aspects of our behavior before the final outcome has been fully realized. When we teach children to swim, for example, we might reinforce them initially for just getting into the water. After the first few lessons, though, we do not bother praising them for entering the pool, but save our praise for other signs of advancement, like an improved kick.

Earlier we considered the white rabbit example, in which a little boy lost his fear of a white rabbit, but only gradually. We can compare that form of treatment with "shaping." The gradual process of moving the rabbit closer to the boy is called "desensitization." Essentially, we grow complacent about objects in the environment that we see all the time. "Shaping," in contrast, involves reinforcing new behaviors as they unfold. In other words, desensitization involves the loss of fear through the gradual presentation of a feared object, whereas shaping involves the learning of a new behavior by systematically reinforcing parts of the behavior until the whole has been mastered.

From experience, we know that some behaviors are reinforced on a regular basis and others are not. If we are salaried employees, our checks come on a predictable basis. When reinforcement occurs on a regular schedule, it is called a fixed schedule, of which there are two types: fixed interval and fixed ratio. A fixed interval means that we get paid every two weeks (or once a month, etc.). A fixed ratio might apply to a sales job in which we get paid $10 for every $1,000 of sales.

In the real world, we may be unable to predict when we will get paid. This might happen if we're starting our own business and do not know when the money will come in. This type of reinforcement schedule is called "intermittent reinforcement." (All this may seem tedious or technical, but information about reinforcers can be very helpful for changing behaviors or habits.)

As it turns out, a lot of research has been conducted on reinforcement schedules, with an important finding. Intermittent reinforcement is far more effective for keeping a behavior going than other schedules of rein-forcement, although this assertion may seem illogical.[27] Here is an exam-ple to help us understand why. When people drop money into slot ma-chines often enough, sooner or later they are "reinforced" on an intermittent basis. If gamblers received a payoff every single time, and then the slot machine suddenly stopped and never paid again, people would quickly learn that the proverbial well had run dry. If the machine only pays inconsistently from the beginning, though, people will keep playing much longer because they never know when it will pay off again. Humans and animals are far more likely to maintain behavior longer on an intermittent reinforcement schedule than other schedules of reinforcement.

When I was a new professional in the mental health field, our office often received referrals from local physicians. I met one woman who had been physically abused by her husband on eight occasions, yet she re-mained quite attached to him. As I got to know her better, I learned that his attacks were often, but not always, followed by acts of "contrition," which included flowers, apologies, and dinner at a fancy restaurant. His behavior was "intermittently reinforcing" (it reinforced her for remaining with him). Intermittent reinforcement is a very powerful means of maintaining a behavior. (I should add, the whole issue of domestic violence is compli-cated, in terms of motives for remaining in an abusive relationship, but the dynamics of "intermittent reinforcement" often play a key role.)

Avoidance Learning.   There is yet another type of conditioning that follows many of the same principles we have been discussing. We learn to avoid or escape unpleasant events, and we do so all the time. This is one of nature's survival mechanisms, and if we did not learn to avoid threat or danger, we might not survive.

We can learn to "anticipate" negative situations and avoid them before they occur. Even animals do this, which is why I put the word "anticipate" in quotes, so as not to imply a conscious, reasoned connection. For example, it is not uncommon for people to housebreak a dog by using a rolled-up newspaper. Fido may quickly learn to go outside when nature calls, but he may also learn something else. The sound of a newspaper teaches him that it is time to leave, never mind that we are just reading the paper.

Some of the factors that influence the effectiveness of punishment also influence the effectiveness of "avoidance" learning.[28] First, we learn to avoid events more quickly if events are *really* unpleasant. If we absolutely detest large groups of people, it only takes one surprise party before we will develop a keen strategy for not being "surprised" again. Second, as in the case of punishment, the longer the elapsed time between some unpleasant consequence and a behavior, the *less likely* we are to develop an "avoidant" behavior. Timing is everything.

Although necessary for survival, there is one aspect of avoidance learning that can cause problems, not solve them. If we chronically avoid situations that cause us fear, our propensity to avoid can become highly ingrained. If we were badly embarrassed by walking into a kindergarten class at the age of five, for example, thereafter we might avoid entering rooms full of people. As adults, however, it is likely that we have improved coping skills in comparison to when we were five, but if we do not try a new behavior, we will never know.

Brain Systems. Earlier we discussed one criticism of psychology—its failure in some instances to include the brain in many theories relating to behavior. Nevertheless, as we have seen, behavioral psychologists have discovered techniques that are highly effective for changing behavior, and as it turns out, one reason that behavioral strategies work so well is because those strategies may mirror internal brain processes. In other words, behavioral learning principles may be effective because they duplicate the same principles that operate *inside* the brain.

At the cellular level, even neurons appear to follow many, or all, of the principles of conditioning that we have been discussing.[29] That is, neurons can become "sensitized" (more reactive) when they are used; their behavior can be "extinguished" (less active) when they are not used.

Similarly, studies on individuals with obsessive compulsive disorders are enlightening in this regard. *Obsessive compulsive disorders* generally involve unwanted thoughts or behaviors that individuals experience as being out of their control. Someone might feel compelled to check the locks on his door several times each morning before leaving for the office, to the point where he is often late because he returns home to check the locks when he gets a few blocks from his home.

Research on the brain and behavior has shown that these types of habits can be eliminated by following the behavioral principle outlined in this chapter. PET scans have demonstrated that behavior changes result in measurable changes within the brain.[30] It seems, then, that changing behavior is a direct means of changing the brain, and by following the types of principles outlined here, we have powerful tools for changing the brain and behavior.

Clearly, scientists have learned a great deal about various conditioning systems, as we have seen, and have identified various brain areas and structures involved in *some* forms of conditioning. As we move up the nervous system, both figuratively and literally (if we are standing), we move into the area of the limbic system. As mentioned earlier in this chapter, one of the key players in that system is the amygdala.[31] The amygdala is the brain structure most often implicated in the experience of fear. Fear, in turn, is the mechanisms that underscores avoidance learning. (Incidentally, long-term fear memories are not stored in the amygdala, which reiterates another point: Learning and memories can be moved from where they are first manufactured.)

Experts disagree with one another about the extent to which conditioned learning is influenced by conscious thought processes, but at least some experts believe that complicated events are processed simultaneously by different regions or pathways within the brain.[32] Thus, operant conditioning in humans may involve many regions or structures within the brain, so at present, it is not possible to pin down accurately all the brain structures that are essential for this type of learning.

We now have our next general principle arising from this chapter.

PRINCIPLE 9: THE OPERANT (LEARNING) PRINCIPLE

Operant conditioning is another type of learning. It, too, can occur outside of our awareness. By definition, this type of learning is influenced by the consequences of our behavior; that is, operant learning is based on reinforcement and punishment.

The role of conscious processes in human operant conditioning will remain a point of debate for many years, but there is little doubt that even "reinforced" behavior can be modified with or without someone's awareness (or consent).

I acknowledge that the idea of being "conditioned" out of our own awareness or against our will is repugnant, raising concerns about "Big Brother." However, in my view, we are better off knowing this information than being oblivious to it.

To illustrate how people can be influenced through reinforcement, I offer an example from my undergraduate college experience. I wish to emphasize, this example is not intended to illustrate "proper behavior," only to illustrate a point. I was enrolled in a sociology class. One day, everyone in the class was approached before class, individually, by a student who claimed to be conducting his own sociology experiment. We were asked to look slightly to the side of the professor as he lectured, rather than giving him direct eye contact. Gradually, the professor moved over to gain our eye contact. We would then give him eye contact, for a while. Then, we shifted our gaze slightly to the side again. Once again, he moved over. Eventually, we had him backed into the corner just by shifting our eye contact (by manipulating reinforcement).

I presented this example only to demonstrate that we are all subject to the principles of conditioning, for better or worse. By taking control of these principles, though, we can put them to positive uses to benefit ourselves and others. I should also add, professional psychologists have strict codes of professional ethics, and they are not allowed to practice psychology on individuals who have not given their informed consent to be treated.

If I personally had any concerns, which I do, it is about the unscrupulous use of behavioral principles. However, perhaps the best way we can protect ourselves from the unethical use of these techniques is to know them well.

IMPLICATIONS FOR UNDERSTANDING BEHAVIOR

In earlier chapters, I may have left the impression that once the brain was fully developed, and once memories had been formed, each of us was stuck with an unalterable fate. Although none of us can go back and redo an event that has already occurred, by using some of the principles presented here, we can make any number of changes for a better future. There are so many ways to change and influence behavior, and many of those ways rely on the principles just highlighted. Here are some ways of putting our knowledge to use.

Using Operant Conditioning Principles

With a little understanding of conditioning and reinforcement, we can make enormous changes to our own behavior, which also affects our brains through learning. The most important key is the knowledge that *most* behaviors are maintained through principles relating to reinforcement. In other words, the first step to behavior change, and therefore brain change, is accepting the idea that, as far as we know, *all* behavior unfolds for a reason.

"Behavior modification," as these principles are often labeled, sounds clinical and dry. I no longer think of these principles in that light, however, because I have come to see how useful they are, and most people, in my experience, are relieved to discover techniques that help them eliminate unwanted habits or learn new behaviors that enhance the quality of their lives. Using a school setting as an example, we can look at ways of putting these principles to practical use, but we will also learn a different way of looking at the concept of reinforcement.

To identify "natural" reinforcers, we only need to look at what happens immediately after any behavior has occurred. If a student is making wisecracks in class, we would want to observe how other students react to the "wisecracker." Are they laughing? Are they responding in other ways? If so, both the motive and payoff of the misbehavior is probably "attention."

Reinforcers are essentially the same thing as "motivators" or motive. We are all motivated. If I joke around at a business meeting to be noticed, "being noticed" is an attempt at acknowledgment, recognition, or affiliation (all of which are social needs). Once we have a working hypothesis about "why" people (or ourselves) behave as they do, we can more effectively devise positive ways of changing behavior (*if* that is our goal).

If we consider "reinforcement" to be synonymous with "motive," we can reasonably conclude that the "payoff" (reinforcement) is important to the individual. So, rather than depriving someone of an important outcome, we only need to determine another way for the person to get what he or she needs, but by substituting a *different* behavior.

If a young man jokes in class to receive attention, we can make a deal with him, that for every hour of class work he completes, he can earn five minutes of time during which he can talk to a friend. If he changes his behavior in response to our deal, then we have discovered his motivation. This is just an example, but the point is that by trying to understand why people do what they do (and the reasons are usually biological or social), and by working with individuals to figure out different ways of getting the same need met, unwanted behaviors or misbehaviors can be changed, often very quickly. If our intervention does not work, then we should consider another possible motivation and devise a new strategy.

When we are trying to change behavior through the use of reinforcers, biological reinforcers are easy to spot. A child squirms in his seat when he needs to go to the bathroom. If a child consistently loses concentration at 11:30 A.M., hunger is the likely culprit.

It is social reinforcers that we often fail to consider. Why? Partly because what is rewarding for you may not be rewarding for me. So, if we forget about the concept of "rewards," we can think about reinforcers by examining what people get out of their behavior, by looking at the outcome of the behavior. If your son misbehaves after you have given him a warning, look at how you respond to his misbehavior. Again, when we assume that people need or want the reinforcers they earn, we're well on our way to knowledge that can allow us to orchestrate changes in behavior.

Jane is a single mother with a teenaged daughter. Jane noticed that her daughter, who is normally well behaved, recently started to make obnoxious comments, but only when Jane brought friends home. What might the payoff be for Jane's daughter? We can hypothesize that the daughter misses the attention she normally receives when her mother is not entertaining guests. "Attention" (affiliation) is the need.

To test this hypothesis, Jane could make an agreement with her daughter. "If you are on your best behavior when Tyler comes over this Friday, you and I can spend the next day at the mall together." If Jane's daughter does not misbehave when Tyler comes over, our hypothesis has been borne out.

It is perfectly alright for our hypotheses to be mistaken. We do not need to throw out the method just because the hypothesis does not prove correct. We only need to change or modify our hypothesis.

Some other examples might help us figure out ways of changing behavior without removing the reinforcement for the behavior. (The reason for not wanting to remove reinforcement is because behaviors *often* achieve important goals for an individual. If we deprive others, or ourselves, of important goals, we are setting the stage for failure to achieve the change. So, the trick is to figure out ways of getting what we need, but through alternative behaviors.)

Overspending

If we overspend, we should ask ourselves, "What do we get out of this?" (We have to be honest with ourselves if this is to work.) I went to a workshop many years ago in which the facilitator was talking about over-spending. He mentioned that overspending, for him, was a way of "mis-behaving" (even though he was an adult). Whenever he felt "neglected" he went out and bought something. It was a way of appeasing his feelings of "neglect" and a way of making himself feel better. He, therefore, de-vised a novel strategy. He put his pocket change into a piggy bank every day, and on a regular basis, he took the money out and bought himself a present. Interestingly, he found that he did not need to overspend by thousands of dollars to get the same degree of satisfaction he got from spending a handful of coins.

Often, when it comes to motive and reinforcement, we are dealing with nonconscious processes, and a $5 gift may serve the same need, non-consciously, as a $15,000 gift.

Here is another example, based on a client with whom I worked several years back. This man had accumulated thousands of dollars of credit card debt—debt his wife knew nothing about. He had new shirts in his closet that he never wore, yet he could not walk through a store without buying something new, and he always paid with credit. He admitted that he did not wear most of the clothes he bought. Often, he did not like his purchases. He seemed genuinely perplexed by his own behavior.

As I came to know him, I learned that his wife had a drinking prob-lem. He also told me that he came from an "alcoholic" family. Eventually, we formed an hypothesis. He hated his wife's drinking, but never told her. Whenever she drank too much, he became angry, and shortly afterward, he went on a spending spree. It was a very indirect way of expressing his anger. (That was our hypothesis.) So, what did we do? The payoff (rein-forcement) in this scenario was his "blowing off steam" by spending, which reduced his anger. Since he still needed to express anger, we mod-ified the way he got his need met. He agreed that the next time his wife drank to excess, he would avoid her as much as possible while she was under the influence. When she was sober, he would tell her that he had been (or still is) angry about her drinking, and then he would take a "time-out" by going to a movie. He enjoyed movies, yet seldom went. By going to a movie, he accomplished three goals at once. First, he did not over-spend, as he used to. Second, he was able to get his mind off the conflict temporarily (which is a type of negative reinforcement). Third, by doing something he enjoyed, he was "reinforcing" a new behavior, that of telling his wife about his anger. Very quickly, he stopped spending so carelessly, and he began a new pattern of addressing the issues of his own anger.

Overeating

Earlier, we mentioned Jack, the 34-year-old salesman who reinforced himself on a continuous reinforcement schedule, dropping in at his favorite donut shop whenever he wanted. The payoff is ostensibly the donuts, but first, we should ask, "What is there about the donuts that he wants or needs?" Is it the taste? Is it the feeling of fulfillment that comes from feeling full? It could even be the conversation he has with other patrons while he is at the shop.

If we are going to change a behavior, we need to decide first what need is being met. Assume that Jack eats because he likes to feel full. Perhaps he grew up in an environment of poverty in which he never felt completely full. So, if he wants to lose some weight, we figure out new ways for him to feel full, but through substituting another behavior that also allows him to experience the feeling of being full. We might, for example, ask him to drink a quart of water every time he gets the urge for donuts, and then go ahead and drive to the donut shop. By the time he gets there, he will feel full, and possibly eat less. Since primary reinforcers are, by definition, something that people need, a strategy to change behavior is going to fail if a reward is taken away without attending to the underlying need that is being fulfilled.

Social Problems

Speaking strictly from the point of view of conditioning, one reason why so many social ills are difficult to remedy is because problems are caused, collectively, by behaviors that are too distantly removed in time from the consequences of careless actions. If someone throws a plastic foam cup into a city lake, for example, the possible adverse consequences to the individual may be too distantly removed in time for the person to make a connection between the behavior and its consequences, so the behavior has a low probability of being altered. This general concept is particularly critical for understanding large groups of people and the effects of their behavior over time.

Power of Facing Fear

We often develop strong adverse or positive emotional reactions to other people or places, yet, many of us feel unable to change those feelings through any conscious means. Many people fear heights. The neocortex does not override these types of fears, although it may give us awareness that we are afraid. "I know that I am terrified of heights, and I know why." The knowledge of why we are afraid does not dislodge the fear, but we can use conscious means to *indirectly* modify a fear.

One of the principal characteristics of classical conditioning is that once a conditioned association is broken, it is often broken forever. Thus, even if we are deathly afraid of heights, if we can manage to get into a height situation, possibly with the help of a trusted friend, *and stay there* long enough, sooner or later our brains and bodies will no longer be able to sustain a high level of fear. This is a general principle, but when we face fear or other strong emotions, they tend to lose some of their grip. The presumed brain mechanism is called "desensitization." In short, if we can ride out the wave of fear and break the fear bond, it is often broken forever. Also, we can use our conscious reasoning abilities to figure out strategies for devising situations that will help us counter our fear, possibly by enlisting the help of friends or professionals who can help.

Graduated Practice

In my bike accident example, described earlier, I could easily devise a strategy to get over my fear of tall curbs if I wanted to. (In truth, I have been unwilling to spend the time because overcoming my fear of curbs is not a high priority for me.) If I wanted to, though, here is how I would do it. I would line up several "curbs" or steps, with each one in the line being slightly taller than the previous one. I would practice riding over a small step; and when I felt thoroughly confident of my ability to ride over that, I would progress to the next taller step; and when I had mastered that, I would go up another notch, and so on. This is a "desensitization" or "counterconditioning" approach, one that would be very effective if I wanted to take the time to do it.

CONCLUSION

Humans and animals alike can be conditioned. Conditioning, as a form of learning, generally requires that a connection is made between a behavior and some outcome. The connection, however, does not have to be consciously known. The same conditioning principles that work on animals also apply to humans. Behavioral psychologists have discovered a whole host of learning principles that have enormous utility for both understanding, changing, and predicting behavior.

This chapter has also produced two of the twelve principles for understanding behavior, restated here.

PRINCIPLE 8: THE CLASSICAL CONDITIONING (LEARNING) PRINCIPLE

Classical conditioning is a type of learning that can occur outside our awareness, and one particular form of classical conditioning supports "emotional learning." As a type of learning, classical conditioning cannot be altered through reasoning, but it can be changed through specialized "counterconditioning" procedures.

PRINCIPLE 9: THE OPERANT (LEARNING) PRINCIPLE

Operant conditioning is another type of learning. It, too, can occur outside of our awareness. By definition, this type of learning is influenced by the consequences of our behavior; that is, operant learning is based on reinforcement and punishment.

* * *

Conditioning is a class of learning that is critical for survival, but it is not the only type of learning that keeps us healthy and safe. The next chapter will explore another type of ancient learning, one that we also share in common with animals. It is a learning system that may be every bit as important for our survival as the conditioning systems we have been discussing.

CHAPTER 8

SEEING OR HEARING
IS DOING

Two cheetah cubs watch their mother closely as she chases down an antelope, and shortly afterward, the cubs awkwardly mimic their mother's behavior while chasing butterflies.

A five-year-old girl learns a new word while watching her older brother struggle with a flat tire. Later the girl repeats the new word during a holiday dinner with all the relatives present, much to the older brother's embarrassment!

A six-year-old is spanked by his father for carelessly running into the street, and later that day, the six-year-old hits his younger sister for "misbehaving."

SOCIAL LEARNING

Each of the above examples demonstrates a type of learning that, like conditioning, has its origins long before the first humans ever walked the earth. *Social learning* involves the type of learning that comes from watching or listening to members of our own species—the learning that teaches us to successfully live among humans.

To many psychologists, social learning refers to a specific doctrine associated with the respected psychologist Albert Bandura of Stanford University.[1] However, as used here, the term "social learning" is being employed more broadly to refer to many forms of learning that occur simply through contact with other people. Other expressions that refer to this type of learning are "observational learning" or "imitational learning."[2]

Some people might argue that most learning depends on contact with others. Even formal education places students in the presence of teachers.

However, by social learning, we are considering only the type of learning that teaches young humans how to live in a social group; *the type of learning that appears to unfold naturally, often without conscious awareness*. For example, as our vocabularies were expanding rapidly during the third, fourth, and fifth years of life, we were not purposely trying to learn new words. Our knowledge expanded rapidly just by being around others who were speaking.

The topic of social learning has even become a national priority, as illustrated by the following quote taken from a publication of the U.S. Department of Health and Human Services:

> Because humans are social animals, understanding how people perceive and respond to the social world is one of the most important tasks of basic behavioral science research. Many of our Nation's most pressing problems—including mental illness, drug abuse, violence, and social and racial conflict—reflect disruptions and distortions of these fundamental social processes.[3]

When it comes to both individual and group survival, social learning is every bit as important as other types of learning, maybe more so, because it is the learning that teaches us how to live within society. It is one of our most fundamental learning systems because it underlies so many cultural lessons.

Much of our learning about appropriate social and communication rules occurs completely outside of our awareness. We construct complicated sentences every day, but even the most highly educated linguists find it challenging to enumerate all the rules of grammar and syntax we follow when articulating even a simple sentence.

We previously explored the concepts of "critical" and "sensitive" periods. A great deal of social learning appears to occur during sensitive or critical time frames when the brain is most receptive to important lessons. However, even older humans learn throughout life by observing others, so it is not accurate to conclude that observational learning is invariably dependent on sensitive time frames.

The important point is that social learning is acquired with little apparent effort on the part of the learner. Further, in contrast to operant conditioning, social learning may even take root in the absence of reinforcement, although this is a controversial point. Without debate, however, social learning, like classical conditioning, is an extremely efficient form of learning, and many lessons so imparted will remain with us for life.

When I was quite young, I remember my mother telling me that "sugar draws ants." It is one of those lessons that I heard once and never forgot. At the time, the lesson was instantly emblazoned in my memory and imagination. Because of my age, however, I did not have a variety of definitions for the word "draw." To me, it was synonymous with the word "sketch." As absurd as the notion now seems, "sugar sketching ants," I nevertheless believed what my mother said. This story also illustrates another important point. Social learning does not require "rational" analysis or conscious scrutiny. It can involve the assimilation of images, sounds, or ideas with no critical screening. "Seeing is doing." "Hearing is believing."

What we see and experience, especially during childhood, has far-reaching ramifications for our "understanding" of appropriate social behavior. This is a type of learning that is not often analyzed by the child's mind, partly because neither the neocortex nor the corpus callosum, which connects the two cerebral hemispheres, is fully developed in early life.[4] This latter point is important because some forms of emotional conditioning may arise from the right hemisphere, and when the corpus callosum is undeveloped, it is possible for emotional conditioning to influence behavior outside of an individual's awareness.[5] Thus, many lessons imparted through social learning are absorbed, in toto, without critical analysis.

From the perspective of adaptation and natural selection, we would expect to find a learning system that incorporates complex behaviors quickly and efficiently. As we know, the rules we follow when in the presence of others are often very complex, and we would be hard-pressed to teach those rules in words. Fortunately, children learn these rules quickly by watching what others do, which is a mixed blessing.

Studies on aggression provide further insight into social learning. If children are physically punished, they will often punish other children. If children are verbally reprimanded, they will often reprimand their peers. If children watch violent television programming, they will react with violence. Finally, to confirm what we have been discussing, parents who were abused as children often grow up to abuse their own children.[6] Thus, in spite of its efficiency, there is another lesson for us to consider. Unlike formal classroom instruction that is intentionally directed, the impact of social learning can be unintended.

Being a social creature for us is a two-sided coin. Some lessons occur rather effortlessly (if we are in the right place at the right time), and fortunately, most of us are exposed to the right lessons with long-standing, positive results. Yet as we have seen, some children may be exposed to environments that are not healthy. Thus, one important corollary to being a social creature is that, while in the presence of others, if our brains are receptive as a result of our age, we cannot remain *uninfluenced*. When the brain is ripe for learning, *learning will occur*, regardless of the lesson that is presented. A look at this principle, in light of a cultural perspective, is needed.

Social learning obviously underlies many of the most important lessons upon which our society rests—lessons of culture, religion, and language. While growing up, we observe how our parents behave in a number of different settings. Dad acts one way at home, but he behaves very differently at the bowling alley when he is with his buddies. Dad behaves in a completely different manner in front of *his* dad than in our presence. Mom, too, appears to have different sets of rules. When she is talking to her best friend she seems more relaxed than when she is talking to the mayor's wife at a fund-raising luncheon. We observe these interactions constantly. No one sits down and formally teaches us the rules of social engagement (for the most part), unless we have committed a gaffe or embarrassed our parents publicly. Rules of culture are learned from immersion, not instruction.

Perhaps one of nature's greatest ironies is that social learning is primitive and literal in many respects, yet it is the type of learning that has traditionally held societies together. What do I mean by "primitive" and "literal"? Social learning is primitive to the extent that we share this learning system with other animals. All social animals, like apes or lions, learn from watching other members of their species. Social learning is "literal" because we absorb lessons and rules that have no particular logic. We cannot go into a laboratory and "prove" that it is "right" to greet a stranger by reaching out to the stranger's right hand. Why not grab his elbow or reach out to his ear? By the same token, we can offer no scientific evidence that it is "best" for adjectives to come before nouns, nor can we *prove* that it is rude to yawn when meeting the future in-laws for the first time. We cannot even prove that one religion is right and another is wrong, even if they espouse opposite beliefs about the role of marriage in society.

Rules governing social interaction are frequently arbitrary and make "sense" only to those who have learned those rules within a cultural context beginning in childhood. Our customs probably seem strange to outsiders, but we learn the rules of our culture without questioning them so they seem perfectly natural to us.

We have considered the assertion that we are social creatures, but it is difficult to explain what, exactly, is meant by that assertion. Nevertheless, coming to some understanding about being a "social animal" is critical if we are to understand ourselves and our brains. Thus, we will begin by revisiting our prehistoric roots.

Evolutionary History

In comparison with many different species, primates are among the most social of all mammals, and our social behavior is assumed to have been naturally selected for its survival advantages.

Although we may take our safety for granted in a modern world, where we are protected by civilization and government-supported police forces, when our distant ancestors walked the earth, they had no such protections. At best, they had only crude weapons, and our most distant ancestors did not have the use of fire. Further, our ancestors had no claws, and they were not particularly fast, compared with many other animals, some of which were eager to eat our ancestors when the opportunity arose. In effect, our distant cousins were sitting ducks, *except* for their superior intellect, which, in part, allowed them to plan ahead and *cooperate with one another in small groups.*

Group

The very heart of any society is its people, and people together compose a group, but not all groups of people exert the same influence over us. Why is that? To answer this question, we should consider the characteristics of a group.

Most obviously, a group must consist of more than one person. Beyond this, however, social scientists tell us there are other conditions that must also be met. First, group members must *interact* with one another and consider themselves to be part of a group. Second, group members must demonstrate a "mutual dependency." They work together to achieve mutual goals. Third, interactions between group members must unfold according to some sort of structure.[7]

In contrast to a group, mere collections of individuals who do not meet the foregoing criteria are called "aggregates."[8] Strangers who are simply sharing the dryness of a bus stop do not meet the criteria of a group (unless the same people have met at the same bus stop on a regular basis, in which case they *might* constitute a group).

The reason we are focusing on the qualities of a group is that we *act* and *think* differently in the presence of others than we do when alone. To illustrate this point, we can examine one of the most famous of all studies ever conducted in the field of sociology or social psychology, "The Robber's Cave" study.[9]

During the 1950s, two researchers, Muzafer and Carolyn Sherif conducted research into social conflict. At a summer camp for boys, at Robber's Cave Park, in Oklahoma, the researchers first set about causing conflict. They gave two different groups of boys different group names. Very quickly, the boys came to identify with their own group. Next, the researchers wanted to test the effects of competition, and they soon learned that competition quickly turned the two groups against each other.[10] "Normal," "happy" boys, as they were described, quickly became aggressive and hostile, but the researchers concluded that "[t]he critical trigger for open hostility was the need to compete for limited resources (prizes)."[10] This study seems particularly relevant because scarce resources, or *the impression* that resources are scarce, appears to be a key factor relating to increased hostility.

We have considered the effects of competition in the face of limited resources, and we have considered the issue of stereotypes before, so we can now look at a closely related issue, and see how social sciences have analyzed it. Several decades of research have focused on desegregation and have studied what happens when people are *forced* to interact with one another, or are in the presence of others whom they would not willingly associate.

In the early days of desegregation and forced busing, it was assumed that greater contact between people might foster improved understanding, but that was not invariably the case. Now, several years later, researchers have discovered what factors appear to draw groups closer together when those groups have been brought together artificially. Forced contact between groups is most effective for *reducing* conflict under the following conditions: First, people must come together in small groups where each member has "equal status." Second, people must "interact" with one another. Interaction, however, helps to reduce tension only if some other conditions are met: Interactions must occur between individuals who are thought to be "typical" of their group.[11] In other words, people will not abandon their prejudices if a minority group member does not appear to be typical of that minority group. Also, interactions must provide the chance for individuals to learn about negative stereotypes and "disconfirm" those attitudes. People have to be given a chance to see the inaccuracy of their stereotypes. Finally, the group must have "norms of equality."[12] They must cooperatively work toward understanding. In the final analysis, people must be motivated to break down harmful stereotypes.

In the Robber's Cave study, the conflict between groups was eventually resolved through the introduction of what was termed a "superordinate goal."[13] This is a goal in which neither group can "win" without the help and cooperation of the other group; that is, when a mutual goal cannot be obtained through any means but cooperation.

We can now apply what we have learned about group involvement to ourselves and our prehistoric ancestors. Clearly, prehistoric hunting and gathering bands met the criteria we have seen for a "group." They interacted with one another, and with little doubt they must have considered themselves to be part of the band. In fact, if anything, some of our prehistoric relatives may not have known that other bands even existed. Although hypothetical, it is unimaginable to presume that they did *not* have a mutual dependency on one another. Likewise, they almost certainly worked together to achieve mutual goals. Finally, it is equally unthinkable that they did *not* have a consistent group structure.

We could even argue that their overriding concern for survival could be depicted as a "superordinate goal," like that described in the Robber's Cave study. This is important because it lends support to the hypothesis that they were egalitarian, that they cooperated with one another. If they did not cooperate, they would not have survived.

In contrast to our prehistoric ancestors, we modern humans belong to many groups, and almost certainly spend more of our time in "aggregates." Shortly, we will draw some conclusions about the possible effects of multiple group membership, but first, we will look more closely at what is perhaps the most basic group in today's society: the family. Since the family is considered to be the primary source of socialization for modern children, we will review some studies that have focused on early life within this type of group.

Early Human Childhood

One way of depicting childhood is to describe it as those years in which humans are dependent on their caretakers. We have seen that compared with other primates, human infants are dependent on adults for much longer in terms of actual years of dependency, and also longer proportionately. That is, the portion of our lives spent in dependency on our parents or other adult caretakers is longer than for other primates. In this society, as we know, it is not uncommon for children to remain at home until the completion of high school, or longer. (Clearly, as children mature, they are not fully dependent on their parents, but a partial reliance on parents typically remains for many years.) Thus, if children remain at home until the age of eighteen, and subsequently lived to be eighty years old, they will have spent about one-fifth of their lives being "dependent" on their parents.

Naturally, some children leave home much earlier, but it is unlikely that they could survive completely on their own without some adult supervision prior to their early teens. Even in the case of children who leave home as early as possible, the years of dependency are still likely to have been many. In the case of a boy who left home at twelve, if he was fortunate enough to survive to eighty, he would still have spent a little over one-sixth of his life in dependency.

In either of the foregoing "modern" scenarios, it may seem as if children spend a good deal of their lives with parents or families. However, what might seem like "a long period of dependency" to us *may not be long enough*.

Again, it is helpful to reflect on our ancient cousins. When the large human brain emerged in a body that is genetically equivalent to our own, all humans were living in small bands. Even before the brain reached its current size in *Homo sapiens*, our next most recent ancestor, *Homo erectus*, also lived in small bands. There are two important attributes of hunting and gathering bands that we should explore further.

First, groups were estimated to range in size from twenty-five to sixty individuals. Further, they were loosely organized and "knit together by close social ties."[14] Third, just as importantly, they were "egalitarian."[15] In the words of one researcher, "virtually all hunter-gatherer societies have an egalitarian ethic."[16] In all likelihood, the *structure* of band society kept a check on dominant members, with the result that food was shared equally and power was shared among all members. Decisions were, therefore, made by consensus, at least among the adults. In short, it was the "egalitarian" ethic (group pressure) that kept dominant people in line, that also made everyone cooperate, and even provided safeguards for the weak and infirm. These are critical attributes because natural selection not only selects genetically imparted traits, but it can also select adaptive behaviors.

Adaptive behaviors, in turn, share an integral relationship to survival. Behavior also has a close relationship to group size and the number of years of dependency within the group because both influence how children are raised and what children learn, primarily through the mechanisms of social learning.

Because the world's population was much smaller during prehistoric times, and because hunters and gatherers required large expanses of land, it is likely that "defective" bands perished. This is a critical issue because it provided a safeguard for all human and animal inhabitants as a whole insofar as band culture and practices were subjected to the forces of natural selection. The most effective band cultures would have survived longer, and less effective cultures would have died off sooner. "Most effective," in this case, almost certainly applied to how children were raised and what they were taught. Thus, we can tentatively conclude that prehistoric child-rearing practices were subjected to the test of nature. Either children received adequate socialization or the band as a whole perished.

In today's world, the most important social relationship that we have been involved in is that of our first relationship with our caretaker(s), almost always our mothers, and for years, researchers have considered this relationships to be essential for healthy adjustment. As Willard Hartup states, "every neurologically normal baby who has the opportunity manifests a focused attachment." The normal attachment between a mother and child must occur within a relatively narrow window of time and the second six months after birth is critical.[17] Children who do not achieve normal attachment experience devastating consequences. (Previously, we read that this assertion may be in doubt. However, there are so many years of research that support the need for a firm attachment between babies and their mothers that a single study of dissent is more likely to be disproved than the reverse.)

By eighteen months, children who have formed close bonds with their mothers, in comparison with children who remain distant or avoidant, already show major differences from their nonattached counterparts. Well-bonded children are easier to engage in constructive play, they become frustrated less easily, and they show more interactions with their mothers in play activities.[18]

Experience from both the United States and other countries has shown that institutionalized children may show a multitude of problems in social situations later, including difficulty forming relationships, poor peer relationships, and fewer friends than children with past histories of healthy childhood attachments to their mothers.[19]

Studies have shown that, far from being random acts of frustration, or mere signs of discontent, the emotional expressions of newborns are purposeful and goal-directed. In part, early emotional communications influence the behavior of adult caretakers. Long before any of us utter our first words, we are already capable of communicating about our needs through emotional displays. Infants express themselves emotionally, and by about ten weeks, they can even respond to the emotional communications of others. In sum, shortly after birth, infants *actively* engage in *meaningful* two-way communications—primarily through displays of emotion.[20] For many years, researchers believed that "socialization" was a one-way street in which the mother socialized the child, but it is now believed that socialization is a two-way street. Infants, through their emotional displays, exert major influences on their caretakers.[21]

When a mother tries to understand and meet the needs of her infant, she engages in what some have called "empathic communication." Empathy is generally described as a type of communication in which the emotional needs of others are recognized and acknowledged in some way. Empathy may even result in measurable physiological changes in the individual expressing empathy.[22]

When mothers understand their infants' needs, profound, permanent, and positive consequences can result.[23] Specifically, children of these mothers are more likely to be curious and explore their environments. These children appear more cheerful and engaged, and as they grow older, they are more likely to ask for help when needed. In turn, the behaviors have permanent, healthy consequences on both development and *intelligence.*[24]

It should be noted, though, in spite of behavioral and emotional displays as early forms of communication, even the best-intentioned mothers do not achieve anything close to complete understanding of their infants' needs during the first year of children's lives.[25] Stated another way, even the best caretakers can often misjudge the needs of their newborns. Because a baby's emotional expressions are not invariably understood and properly attended to, *all normal infants* develop temporary coping strategies.[26] Children learn to rock themselves, divert their attention from desired objects, suck their thumbs, or hold a blanket. All these behaviors are *normal*, if they are used as temporary coping strategies. Children who learn to handle frustration develop greater confidence and maturity. The ability to self-soothe in the face of temporary frustration is a critical developmental task.

In contrast to well intentioned or empathic mothers, however, some mothers *chronically* ignore their children or consistently misinterpret their children's needs (probably because they, the mothers, were misunderstood as infants). We have all seen mothers or fathers who offered a bottle of milk at the first sign of distress, even when physical contact may have been the baby's primary need.

When a baby's communications are chronically ignored or misunderstood, normal temporary coping strategies are converted into chronic, pathological behaviors. The child looks around less, explores less, responds with lethargy to the environment, and shows less flexibility.[27] (We already learned that neglect *can* result in *permanent* intellectual loss, and that observation precisely coincides with the present discussion on emotional development.)

The brain is experience dependent. So, when temporary coping strategies become permanent, children often turn inward and withdraw. Withdrawal, in turn, curtails curiosity and exploration, which may set the stage for further intellectual decline, and possibly even mild retardation. Although this is a broad generalization, there is growing evidence that profound neglect can lower intelligence, and perhaps the chronic withdrawal seen in some children is one cause.[28]

Positive, early attachment to one's mother or other primary caretakers appears to have lifelong effects. One study looked at the relationships of securely attached infants after those infants had reached adulthood. These individuals were found to be supportive of their spouses, and they generally reported feeling happy. Among adults who had insecure attachments as infants, they experienced discomfort with closeness, and they believed that "true love" did not last.[29] If these findings are accurate, then the consequences for insecurely attached children (later as adults) may be more serious than mere differences of opinion about the lasting nature of relationships.

The ability to reveal our feelings to others has positive health consequences, and the reverse is also the case.[30] People who are unable to share their emotions with others are likely to have more problems related to health. In contrast, those of us who can talk about and reveal our emotions are better off, from a health-risk point of view, than individuals who can talk only about "facts." As an aside, we have all met people who talked nonstop about the most personal details of their lives. Often, though, individuals like this are not in touch with their innermost feelings, so they do not benefit from their constant talk. Also, when people speak too often of deeply personal matters, they can drive others away, which increases isolation and loneliness. In general, though, people who can *appropriately* share their feelings with others have a health edge over those who cannot.

Then and Now. Earlier, we learned that the brain, in particular the frontal lobes, continues to show some signs of maturation, possibly until twenty-one years of age, and the frontal lobes are the part of the brain that have "special significance for human society and culture."[31] Thus, during human prehistory, young adults developed invariably within the confines of small social groups. That is not necessarily the case today where young adulthood might unfold in total isolation, or within another structure. This is important because, as we have seen, the brain is highly dependent on what it is exposed to during development, because social influences shape the brain for later life. Thus, during human prehistory, young adult brains were exposed to elder members of their band for guidance and instruction. In contrast, young adults today *may* pass through a critical developmental phase (young adulthood) in the absence of older adult guidance.

During prehistoric times and continuing to the present, social learning maintains communities, and it may be the *principal* means by which cultural knowledge is passed on to each succeeding generation. When humans lived in small, rather autonomous hunting and gathering groups, important cultural lessons were not subjected to outside or contradictory influences. Those beliefs, however, *were* subjected to "trial by fire." If cultural practices were not adaptive, the entire band could perish.

In today's world, we see many changes from the time of our prehistoric ancestors, relative to the dimensions we have just been considering. First, modern laws and civilization protect a much broader variety of socialization practices. In effect, modern child-rearing practices are not subject to the immediate laws of nature, as was probably the case for our prehistoric ancestors. Civilization has removed families from the immediate consequences of questionable child-rearing practices, relative to the match between those practices and the natural environment.

Second, in the modern world, even the most revered cultural beliefs can be exposed to all sorts of contradictory or competing influences, through the influence of television, radio, movies, video games, "the Net" (the Internet), or even schools. Schools, after all, may not demonstrate single, overriding goals, because schools can be comprised of diverse constituencies. One group of parents may want the school to emphasize "the 3 R's," another may want a greater focus on issues relating to diversity, and still others may want more emphasis on competitive sports.

We have seen that cultural lessons make no *inherent* sense, which is not a problem as long as those lessons are consistent and are not subjected to competing ideas of "right" and "wrong." To illustrate, we can believe in many gods, or just one god, but as long as everyone believes the same thing, the belief system will maintain group cohesiveness. In general, cultural beliefs are the types of "knowledge" that cannot be proved or disproved; rather, validity comes from group consensus and from the adaptive (survival) value of those beliefs.

One of the most pervasive challenges to tradition, in modern society, has come from population growth and the influence of the mass media. From the very beginning of its existence, experts have wondered and written about the potential influence of television on society.[32] From a social learning perspective, we have cause for concern. When children (and adults) are repeatedly exposed to powerful images, they are frequently unable to censure the messages that arise from that imagery. Pictures and sounds can be absorbed literally and stored nondeclaratively, thus influencing behavior and attitudes outside of awareness.

When a young child sees televised images, like a four-wheel-drive vehicle on a rock outcropping in the middle of the Grand Canyon, the child cannot question the "rightness" or appropriateness of that image. Only years later, and only after years of education, *might* the child learn how misplaced the image really is (that, in fact, it is illegal to drive in a wilderness area or off-road in a national park). Unfortunately, images that are stored nondeclaratively (nonconsciously) can influence behavior for life, even when we are unaware of their existence. When several years worth of questionable imagery is stored nonconsciously, we should be concerned. Quite simply, the brain *never* evolved to be exposed to images that are unnatural. In fact, it is not possible to come up with an example in human prehistory of how children might have been exposed to images that were illusory.

From an historical perspective, there is another major difference from then to now. Not only are modern children exposed to greater numbers of out-of-family influences, but they are exposed to greater numbers of *contradictory* sets of social rules. During human prehistory, it is unthinkable that children within a given band would have been given conflicting rules, whereas in modern society, in the course of one week, children can be barraged with many competing messages about normal behavior. A child can be *told* that it is wrong to kill, yet be exposed to hundreds of murders on television, many of which are depicted realistically.

Through social learning, children often "practice" what they watch. Studies have now shown that areas of the brain can be activated by simply watching others perform an action.[33] Furthermore, young infants appear to have an innate tendency to observe others, and by four months of age, babies are attempting to mimic some of the facial expressions of their parents.[34] This suggests an innate tendency for children to mimic what they see, as a basic component of social learning. Facts like these take on much greater significance, therefore, if we consider that some modern children spend more time watching television than interacting with their parents![35] Given that the medium of television excels at creating unnatural images, we must certainly wonder about the long-terms effects on children and society.

In conclusion, we moderns have retained a "prehistoric" brain, but it is being subjected to cultural and social influences for which it may be poorly adapted. Minimally, there is little reason to believe that our brains evolved to contend with the world of today.

Language.   Language acquisition and development is often excluded within a discussion of "social learning," but we will briefly discuss it here because children must be exposed to language during childhood for language to develop normally, and obviously that exposure must come from other people.

Experts disagree about the extent to which language follows other patterns of learning, however. Language, unlike other lessons we learn, appears to have an innate basis or biological predisposition. That is, most children acquire language with "remarkable speed," within about four years, with surprising accuracy, given the complexity of language.[36] Thus, scientists now believe that we are innately "predisposed" to learn language, and this conclusion coincides with what we learned earlier about synaptogenesis.

Furthermore, children with normal hearing and intelligence from all cultures pass through the same language milestones at the same time, in spite of differences between languages. At about six months, children utter their first "babbles"; at about twelve months, they utter single words; and between eighteen and twenty-four months, they are combining words in systematic ways, regardless of the language they speak. Even premature children pass through the same milestones at about the same age, even though they have been exposed to language for a longer period of time.[37] In short, there is little doubt that we are genetically "programmed" to learn language.

Our ability to learn and use language appears to be uniquely human, but we share some of our abilities with other animals. In the case of language, once again, the brain shows its penchant for combining older systems along with more recently evolved ones, and integrating capabilities into functions that become uniquely human. We share our ability to differentiate the sounds of "b" from "p," or "d" from "t" with both minks and monkeys.[38] Again, this suggests that our ability to make these distinctions is both genetically endowed and comes from older capabilities that predate language.

In sum, language has evolutionary roots that predate humans, but written and spoken symbolic expressions are uniquely human. Our language abilities are endowed through social contact with others; and conversely, language skills cannot develop in the absence of immersion in human cultures.

Brain Studies.   It is not possible to point to a specific area within the brain, and say, "This is the area that makes us human." Yet, attributes like language and social interactions with other humans are defining characteristics of our humanity, and certain areas of the brain are more involved in these types of attributes than others. (I should emphasize that we are speaking in broad generalities here. What would language be like without respiration, for example? Does language make us more human than breathing?)

The human ability to learn from others may be far more than a capability, it is probably a biological imperative. Human infants, as we learned, imitate facial expressions of adults within the first few weeks of life. This ability appears to be inborn and is considered by some experts to be a necessary precursor for social living.

Through the use of PET scans and other technology, scientists have learned a great deal about what areas of the brain are involved in those general processes that contribute to our "social nature."

Frontal Lobes.   The front of the human cortex is a large area, comprising about one-third of the cerebral hemispheres. This area contributes to many functions.[39] For the types of behaviors we have been discussing, the most critical section of the frontal lobes is termed the *prefrontal cortex*. This is an extremely important part of the brain for both humans and primates and is a part of the brain that has grown larger and more specialized in humans than in any other animal.

Although people with frontal lobe damage often score average on standard intelligence tests (they often have an average knowledge of general facts), they can still be grossly impaired in how they utilize their knowledge. Their behavior can appear "fragmented," it may not be goal-directed, or it might be excessively emotional or impulsive.[40]

I reviewed the records of an adult female patient who had sustained frontal lobe damages in a fall. Her measured IQ was average, but on a daily basis, she could not survive without the help of others. She would get lost if she walked into an unfamiliar neighborhood. She would show poor judgment in social situations, often dressing inappropriately and showing a lack of restraint in sexual matters. She frequently took rides from strangers. Finally, if she had any money, she spent it carelessly. In general, people or animals who have sustained frontal lobe damage may lose their ability to function within social settings.

It is believed that the prefrontal cortex integrates incoming data, including sights, sounds, and smells, and allows us to notice constantly changing nuances within our surroundings. If the prefrontal cortex is injured, or fails to develop, an individual is unable to comprehend and respond to subtle cues.[41]

Behavior in social situations is exquisitely dependent on noticing what is going on around us and making constant adjustments to those fluctuating changes. Just about any behavior can be appropriate at the right time. Belching loudly in the privacy of one's home is not a faux pas. At a formal dinner party, though, with the mayor and city council present, it is a fairly major gaffe.

It may seem ironic, but people with frontal lobe damage are not totally *unmindful* of rules, rather, they often behave *too consistently* from one setting to the next. They fail to adjust to different "rules" from one situation to the next. They fail to appreciate the simple fact that there are different standards for waiting in a doctor's office compared with the standards that govern behavior while watching a football game on television in the privacy of the home. The client I discussed earlier would often walk into a waiting room full of patients and show no ability to modify her behavior from how she acted outside. She would dress inappropriately for the circumstances. She would yell or talk very loudly and often approach complete strangers and begin talking to them as if they were long-lost relatives.

The frontal lobes also play a key role in assimilating incoming sensory and emotional data, so they help moderate and integrate emotional "information" from the limbic system. Damage to the frontal lobes or failures of the lobes to develop normally can cause individuals to behave oddly in social situations, miss important social cues, or respond emotionally in ways that are inappropriate.[42]

In addition to the frontal lobes of the neocortex, particularly the prefrontal cortex, the amygdala within the limbic system is also crucial for appropriate social functioning, especially for appropriate emotional displays in social settings. Whether it is a normal fear response or a normal expression of anger, the limbic system is heavily involved in appropriate behaviors.

Now, in light of what we have been thinking about, we can formulate the next major principle for understanding behavior.

PRINCIPLE 10: THE SOCIAL PRINCIPLE

During infancy, we humans are completely dependent on others for our survival, and during our prolonged childhood, we readily learn language, many cultural rules, and the basics of living in a society—merely through being in the presence of others. However, during the earliest years of our lives, we cannot resist the influences of other people, whether those influences are positive or negative. Finally, even as adults, good health and mature functioning are heavily dependent on positive social interactions with others.

As we have seen, social influences on us are ubiquitous. Socialization is a two-way street, however. Initially, we cannot live without the help of others, and when in the presence of people, we cannot *not* learn from them. This principle speaks to the importance of healthy environments, both physical and social, and with this knowledge, we can exert profound and positive effects on ourselves and others.

IMPLICATIONS FOR UNDERSTANDING BEHAVIOR

Whereas facts and information have to be encoded before storage occurs, visual imagery, especially for the young, cannot be critically analyzed. The young brain does not have the analytical skills to question such "information" (and visual imagery is probably stored differently from "factual" information). The evidence for this comes from the fact that humans mimic what they see and often mimic what they hear, especially during childhood. Although social learning is especially critical among the young, all of us can learn throughout life, and one of the best ways is through watching or imagining the actions of positive role models.

Children (and adults) can learn many new behaviors by observing individuals who do well on some task and then mimicking the actions of those people. In short, the human ability to learn from watching others is one of our most powerful learning systems, one that remains available to us for life. With this knowledge, therefore, we can exert a tremendous influence over our own behavior, if we choose, often by our selection of or influence on the groups to which we belong.

Healthy Groups, Healthy Minds

The power of groups to influence our behavior cannot be overestimated. We can use the power of groups to accomplish many of our goals in life, from losing weight to making our neighborhoods safer places in which to live. Most of us are subject to the influences of groups, even if we try to resist those influences, so we might as well capitalize on the power of groups and either form positive groups or modify the ones in which we are already involved.

Years back I worked with groups of men who had been court-ordered to attend therapy because of "domestic violence." They had physically transgressed against members of their families, usually their wives or female partners. Upon first meeting the potential member during an initial interview, each was invariably angry, hostile, and resistant to treatment, especially since group treatment had been court-ordered. However, once the men came together as a group, and once the group had met several times, the combined influences of all the members (the group) came to exert profound influences over each member's behavior. Even the most belligerent and rude individuals came to act respectfully and politely, and all of them learned to talk about their feelings to a much greater extent than they had been able to early on. I am not suggesting any of them were "cured," per se, nor do I know if their improvement extended beyond the group. However, the power of a group over its members is profound.

There is yet another way we can use groups to help ourselves or others. Because most of us have deeply rooted social needs, we can rely on groups as support systems to help reinforce changes we are trying to make. Many self-help groups are effective, in part, because they help to reinforce changes that members are trying to make. Members can tell the group what they wish to accomplish, and then rely on the group for support and encouragement. Even if we consciously put together a group of people and tell them what we need from them, the group can still exert enormous influences over our behavior.

People who are not influenced by groups are often those who suffer from genuine, diagnosable mental illnesses. One of the better examples that received wide publicity was that of Ted Kaczynski, the Unabomber, who was diagnosed with schizophrenia.[43] Often, people with schizophrenia are unable to tolerate the presence of others, and for many, even a therapeutic group is intolerable. Thus, many remain cut-off from others and therefore remain uninfluenced by group pressure. This is important because it attests to the close relationship between healthy individual functioning and positive functioning within a social setting. (As an aside, since I mentioned Kaczynski, it should be stressed that most schizophrenics do not harm other people.)

Naturally, not all groups provide a wholesome environment. We have all heard about crime syndicates or youth gangs that exerted tremendous, yet negative, influences over their membership. We often hear that learning to resist "negative" group influences is an important social skill. However, it is highly unlikely that our brains evolved to accommodate this challenge. It is far easier to resist negative group pressures by not associating with negative groups than by trying to resist unwholesome influences from within a damaging group (which may not be possible for many people).

Sharing the Wealth

We have already read about the advantages that social living bestows on us all. However, there is another very obvious benefit of social living, a benefit that any of us can put to use at any time. We can learn from the most gifted and talented members of our society. Through classes and private lessons, we can benefit from the best minds around us, and we do not have to be particularly creative ourselves. This is a major advantage we moderns have over our prehistoric ancestors and over other social animals.

In a similar manner, we can use what we know about brain development and groups to influence our own behavior and that of others. The full capacity of the frontal lobes are the last to develop. Such attributes as "wisdom" are also the last to develop. (Wisdom can be thought of as the ability to put knowledge to use in new and novel ways, and the increased ability to consider different perspectives, rather than just seeing the world from one point of view.) We can borrow the wisdom of others by simply being around them. With the power of a group, we do not have to reinvent the wheel every day.

One of society's greatest benefits is that the least skilled of us, in a given area, can benefit from those who are most skilled. We can exert substantial influence on our lives by consciously choosing the people with whom we associate.

Human Interest in Humans

In the first chapters, I raised the question of "why" we humans have such an unquenchable thirst to know more about ourselves. In light of what has been presented in this chapter, I think we can formulate a plausible answer to that question. Based on what we have reviewed, it seems likely that we are interested in people because that interest, in part, was naturally selected in a Darwinian sense. One aspect of being a social being, therefore, might be biological. We may have biological and genetic imperatives that cause us to observe others and maybe even to reflect upon ourselves.

In a later discussion, we will see additional evidence of how our thinking and perceptions are highly influenced by others, so it is very likely that our interest in people is one aspect of being a social creature.

Controlling Imagery

A friend of mine told me she gave away her television set. Although I am not recommending this, the fact remains that by controlling the images we expose ourselves and families to, we can exert profound influences on behavior and our feelings. Although televised images *may* have beneficial effects for children under some circumstances, one problem with social learning is that its effects cannot always be anticipated, as evidenced by children who learn to hit others because they had been spanked (when the intended lesson was obviously something entirely different).

CONCLUSION

In this chapter we have explored a fundamental human learning system, social learning, that is integral to our social nature. As we have seen, social learning is highly dependent on the frontal lobes of the brain and the amygdala within the limbic system. The types of lessons that arise from social learning are critical for individual survival and for the survival of society. This type of learning unfolds rather effortlessly, often without our awareness. Thus, in this respect, it can be nonconscious.

There are three fundamental aspects to the social learning principle, which is restated here.

PRINCIPLE 10: THE SOCIAL PRINCIPLE

During infancy, we humans are completely dependent on others for our survival, and during our prolonged childhood, we readily learn language, many cultural rules, and the basics of living in a society—merely through being in the presence of others. However, during the earliest years of our lives, we cannot resist the influences of other people, whether those influences are positive or negative. Finally, even as adults, good health and mature functioning are heavily dependent on positive social interactions with others.

The issue of social learning takes on a new perspective when it is placed in context. Modern humans have the ability and penchant to change their social environments, often dramatically. In fact, we often change our social environment so quickly that researchers are unable to keep up with the pace of change. This is a fundamental issue because there is a close relationship between mental health, physical health, and the groups to which we belong, but those groups are often changing too rapidly to assess their impact.

* * *

As we have seen, there is an intimate relationship between social learning, group influences, and healthy emotional functioning. Thus in the next chapter we will take a closer look at what is meant by "emotional functioning."

CHAPTER 9   THE RELIABLE BRAIN

Pretend that you are taking a walk through a forest, you come around a huge boulder, and there, not twenty feet in front of you, is a very large feline. If you are unarmed, and normal, the emotion engendered by such an encounter would be pretty predictable—*abject terror*! Now, let me ask you, where in your body would you first experience that sensation? In your stomach? Your chest? Before your cerebral cortex can make full meaning of a new situation, much less decide what to do, strong feelings have been triggered, so what are these things called "feelings"? They are part of a larger phenomenon called "emotion."

THE FUNCTIONS OF EMOTIONS

Emotions and behavior share a unique and complicated relationship with each other. Emotions influence behavior. In turn, behavior influences emotion. It is sometimes impossible to make a distinction between an emotion and the ensuing behavior because both can occur simultaneously. To further complicate the picture, it is possible to experience an emotion, but behave in a manner that is unrelated to the emotion. We can be very upset about a job loss and behave as if nothing had happened.

Nico Frijda, a professor of psychology at the University of Amsterdam, is a renowned expert on emotion. He referred to emotional behaviors as "expressive" because they can appear solely for the purpose of expressing emotion.[1] Have you ever observed what happens when a person arrives two seconds too late at a bus stop? Wild gyrations, such as waving a clenched fist, or the prominent display of a particular digit, are not out of the question. These behaviors are so expressive because they serve no

169

apparent goal-directed function, they appear to be internally generated. In contrast, nonemotional behaviors can be described as "instrumental." Instrumental behaviors are goal-directed. When someone walks outside on the coldest day of the year in a bathrobe, the behavior is not viewed as crazy if the act results in the successful retrieval of the morning newspaper. (The distinction between goal-directed and nongoal-directed behavior is not hard and fast. Emotional behaviors can have a hidden goal, even if only to reduce tension.)

Like all behaviors, emotional behaviors can be extremely diverse. They may be stimulating or amusing, like a lover's spat in a fancy restaurant. Emotional behaviors may be subtle, like a suppressed yawn when our host finds yet another set of pictures from last summer's European vacation; or emotional behaviors can be frenzied, like the out-of-control gyrations of football fans after a winning touchdown.

To most of us, the hallmark of emotion is the strong feeling we experience, but all emotional reactions involve much more than just feelings. Feelings are just the tip of the proverbial iceberg.

As we come to understand more about our emotions, we will see that each one appears to serve a specific purpose. Each one provides us with a slightly different message about our relationship to the world. However, I should add, although scientists have discovered a great deal about where in the brain emotional processes arise, they can never be certain why emotions exist. Like all other brain processes, we can only form educated guesses about the real purposes of our emotions. This is an important issue when discussing emotions, because there is so much controversy about their purpose and function. Nevertheless, because emotions are relatively costly to us (they often require large expenditures of energy), scientists now assume that emotions probably evolved to serve many purposes. This issue is being raised here because there are people who think emotions are no more than vestiges from the past, that we would all be better off without any emotions. Most contemporary scientists, though, place our emotions in much higher regard.

As we study the topic of emotion, certain perennial debates arise over and over. One subject concerns whether there are a limited number of "basic" emotions. At this point, a list of basic emotions cannot be definitively drawn for at least two reasons. First, experts do not agree on what ought to be considered "basic." Is anger more basic than jealousy? Is happiness more basic than rage? What complicates the matter even further is that some emotions arise earlier in life than others. Joy precedes pride from a developmental perspective, so is joy more basic than pride?[2]

Second, we will see that each emotion communicates a specific message that cannot be communicated by any other emotion. From this perspective, it would seem that each emotion serves an essential function, but is an "essential" function the same as a "basic" function? The experts disagree about this point.

As we can tell from the anecdote at the beginning of this chapter, we experience emotions as phenomena that happen to us. We do not experience them as conscious, volitional acts.[3] To some extent, we *can* control how we show our emotions or withhold them (and most of us have learned to control what we do in response to them), but we cannot control the first blush of an emotionally triggered feeling because it is a physiological process outside of normal conscious controls.

In a similar vein, emotions are experienced as either pleasant or unpleasant—there is no such thing as a neutral emotion. Furthermore, the experience of emotion is very subjective and personal, and even the "same" emotion can be felt differently from one person to the next. Similarly, the same emotion might be experienced differently from one occasion to the next for the same person. We might recall two occasions when we felt angry, but also recall that our experience of anger was different on each occasion. This might seem odd, at first, but we will soon see why this happens.

When some emotions are extremely strong, like fear, they can interfere with our reasoning and thinking abilities.[4] We *can* become so overcome by fear that we cannot engage in effective solution-focused thinking. This, of course, is the popular stereotype of emotion. In my view, too often emotions are depicted as enemies of rational thought. How many times have we been subjected to one-sided movie depictions of leading ladies who were capable only of screaming, while overly controlled male leads saved the day? In point of fact, the degree and frequency of thinking paralysis depends on such factors as the specific emotion, the emotion's intensity, our general ability to cope with strong feelings, and most important, previous training about emergency or crisis situations. Physicians can remain relatively calm in the face of medical emergencies because they have been trained to do so. Police officers can cope in the face of extreme danger, in part, because of their training.

As humans, we may experience a broader range of emotions than any other species of animal, so when we begin to unravel human emotional functioning, we have taken a giant step toward understanding ourselves and others. As we take a closer look at each hypothesized function of emotion, we can also consider how emotions may have served our prehistoric ancestors in the world in which they lived.

Nature's Signal

Without emotion, our brains would have an enormous task trying to separate sights, sounds, and odors into meaningful and meaningless categories. Imagine trying to make sense of *all* lights, sounds, colors, shapes, smells, changes in temperature, and so on. Fortunately for us, our brains have a system for separating stimuli that have relevance for our needs. Our brains signal us with an emotion. Generally, "positive emotions," like joy, tell us that a match has been made between a current need and something or someone in the environment. "Negative emotions," like fear, tell us to watch out for danger, potential danger, or they signal us that we have lost something important.

As we have seen, the brain draws upon many areas to perform even the simplest tasks. Thus, each specific emotion is tailor-made by the brain for the demands at hand. Each emotion tells us a slightly different story about our relationship to what is going on around us. (Emotions also signal us about internal events, as in the experience of elation after we have worked long and hard at some problem, and realize that a solution has been found.)

The signaling functions of our emotions also alert us to important social cues. Have you ever heard someone say, "I don't really care if I get invited to that party"? Intuitively, we know that when anger is present, the anger is a far better indicator about what the person *really* believes. The failure to be invited to the party is important, and the tip-off is the anger, not the words.

As a therapist, I have been trained to listen, not only to the words of others, but also to their emotional communications, which can be expressed in a number of ways: changes in voice tone, body posture, or facial expression.

Emotion has often been depicted as the "irrational" component of personality, but as signals about what events are important to us, emotions are unsurpassed. In fact, if any function of the brain deserved the term "reliable," emotions, in my view, should win the prize. One of the reasons emotions have received such bad press is that their messages and purposes have often been misinterpreted or misunderstood.

When emotions are relied upon to judge *factual matters*, they are notoriously inaccurate, presumably because they never evolved to judge external reality. Rather, they appear to have evolved to alert us, as explained. An example will help explain how emotions are *misinterpreted*. Assume that you are walking through the city with a couple of friends just in time to see a man race headlong out of a bank. He jumps into a car and the car speeds off. Moments later, a security guard comes running out of the bank with her gun pulled. The guard turns around quickly, and for a brief moment, her gun is pointed at you! Instantly, your blood is infused with adrenalin, your heart begins to race, and your breathing may become shallow. The guard puts her gun down and then runs back into the bank. Moments later, your friends join you, but you are still in a state of panic, and tell everyone that a security guard almost shot you with the largest handgun made.

A strong emotional reaction can change how a situation looks to us. The gun may have appeared much larger than it was, and the amount of danger may have been exaggerated. The limbic system, which is the general area that supports an emotional response, cannot tell us how big something is, nor can it tell us how dangerous a situation is. Rather, the limbic system provides us with an immediate, "orienting response."[5] "Look out! Something's going on here!"

Factors like size and degree of danger, and even the description of the gunman, can only be ascertained by the neocortex. However, because the neocortex must assimilate so much information, especially in a time of crisis, it takes time for the neocortex to integrate and analyze data from various places, including the limbic system. Nevertheless, only the neocortex can determine what type of gun the guard was carrying. In sum, our emotions signal us generally about events that have personal relevance.

Preparation for Action

A major function of emotion appears to be its ability to *prepare* us for action. This feature has been termed "action tendency" or "action readiness."[6] Preparation for action, though, is not the same as action (although the two phenomena are closely linked).

Part of any emotional reaction includes signals from the brain to other parts of the body to prepare us. This "first alert" capability is very fast, but it is based only on a rudimentary analysis of a situation. To better understand the importance of this capability, we can look at two different vignettes. The first describes a situation without an automatic emotional response, and the second describes the same event *with* an emotional response. (The situation is obviously hypothetical, I should add, because it is hard to imagine the situation without an emotional response.)

We are walking down a dark street at night and hear running footsteps approaching quickly from behind. Without emotional capabilities, our reaction would be completely neutral. The only means of considering the situation would be through rational analysis. We might reason that the person from behind us is either a jogger or perhaps a mugger. Our assessment would be based on the time of day, the neighborhood in which we are walking, our beliefs about the safety of that neighborhood, past experience, and a host of other factors. In the absence of an emotional response, we would likely conclude that we could conclude nothing, based on the information we had. We would have to turn around, wait for the runner, and see what happens.

Consider the same vignette, but with an emotional response. When we first hear the footsteps, our whole body would react. Our heart would accelerate to provide an increase of oxygen to our brain and muscles. Our muscles, in turn, would tense. Without our awareness, nonessential bodily functions would temporarily slow down or cease altogether. Finally, our previous train of thought would be put on hold. All of our attention and mental faculties would be riveted to the approaching stranger.

Our emotions quickly place all systems on alert in advance of a final analysis. *If* a response turns out to be needed, we have been prepared for the challenge. This "action readiness" capability is a crucial evolutionary adaptation. If, the "jogger" turned out to be a mugger, we would have far more energy at our disposal to resist a mugging that if we had been caught completely off guard. Clearly, though, there are limitations and trade-offs to this "action tendency."

First, we are stuck with the ensuing feeling. In the latter vignette we might experience overwhelming fear. In turn, our fear *might* be strong enough to prevent us from taking any action. Also, if it turned out that the footsteps were those of a jogger, our fear would have been unnecessary. Overall, though, a general preparedness provides us with a decided advantage over a body that is unprepared. The second trade-off is that emotional reactions provide us with very little objective information about the situation. Our reaction tells us only that the situation has *possible* relevance to our current needs or goals.

One of the most common myths about emotions is that they *compel us* to behave in ways that we would not choose "rationally." However, research has shown that our emotions do not *invariably* cause us to behave in specific ways.[7] We have all heard about cases where the out-of-control boyfriend assaults his girlfriend with the excuse that her flirtations drove him into a jealous rage. This is a common defense, but one that cannot consistently withstand careful scrutiny.

In spite of the assertion that emotions *cause* certain behaviors, we also take for granted the fact that emotional reactions can be modified through training. We routinely provide training to police officers, surgeons, firefighters, ambulance drivers, emergency medical technicians, soldiers, high-rise steel workers, and so on. This leads to the conclusion that people can and do control their emotional behaviors, even in situations that are highly emotionally arousing. (However, as we can see, *training* is often necessary in order for individuals to control their emotional behaviors. People normally are not equally adept at handling strong emotional reactions.)

One reason people claim that emotions drive their behavior is because they have *learned* to behave consistently in response to a particular emotion.[8] Just as we can train a lifeguard to respond semiautomatically to an emergency, any behavior that is constantly relied upon *does* become more automatic. Thus, someone who frequently loses control of his or her temper is more likely to repeat that behavior, just as someone else who frequently maintains control of his or her temper is more likely to repeat that pattern.

In work with many domestic violence offenders, I have seen men who physically abused their partners insist they could not control their actions when provoked. Without argument, many of them had never practiced better ways to behave, partly because many thought that violence was an acceptable solution. However, once they learned about and practiced alternatives, many of them did learn to modify their behavior.

In any complex emotional reaction, there is always a brief period from the onset of a feeling to the start of some behavior, so there is time to think before action is undertaken. Admittedly, some people might strike out when startled, but the reaction provoked by a startle does not invariably escalate into a full battle.

With children, it often seems as if their emotions *cause* certain behaviors. Unquestionably, in the face of strong emotions, some children do lose control. However, their failure to contain their behavior is often the result of a lack of alternatives. They simply have less training and practice at considering or implementing alternative behaviors. Like adults, though, most children can be taught to control their impulses, and even young children learn to suppress aggressive reactions if adults are available to teach alternatives.

Let me clarify an important issue. We seldom have the ability to control how we *feel*. Similarly, none of us can control all the changes that occur in response to a strong emotion. A near miss on the freeway will increase the heart rate in anyone's chest. The physiological responses that accompany an emotional response occur automatically. However, the fact that our hearts speed up does not mean that we have to strike out or take off running or do anything else. Emotions are multifaceted reactions. We cannot control every aspect of those reactions, but we can learn to control specific aspects, especially our behavior. This is important for teaching alternatives to violence.

While on the general subject of inappropriate emotional reactions, the term "emotional disorder" is commonly used by laymen and professionals alike, but the term is not an officially designated psychiatric diagnosis.[9] Although the term is commonly heard, it is difficult to define what it means. Quite often, we refer to people as having emotional disorders when their emotions work just fine. What is more common, perhaps, is that people act on the basis of strong feelings without thinking or reasoning, but from this point of view, the problem is one of poor thinking or problem solving, not one of emotional malfunction, as such.

Emotional Behavior

When an emotionally triggered "action tendency" *is* translated into behavior, as you probably know from experience, that behavior may have tremendous power behind it. This is not a bad thing, necessarily. That energy might save our lives. If we are scared, we can run faster; if we can muster some righteous indignation at the notion of being attacked in our own neighborhood, then a would-be mugger might think twice before trying out his mugging antics on us.

I will never forget a sign of emotion in action that I witnessed many years ago in Yellowstone National Park. At that time, it was still common for tourists to see large numbers of bears throughout the park. As long as there were bears, tourists seemed compelled to perform questionable antics. One woman was walking slowly with a cane. Unfortunately, she took a shortcut between a mother bear and the bear's cubs. Instantly enraged, the mother bear let out a roar and launched herself toward the cubs, not to mention the poor woman. At that exact instant, I saw both the cane and woman take flight in two different directions! The elderly, frail woman ran to her car without the help of her cane. Such was, and is, the power of emotion.

Our First Communication System

"Emotional expression provides a powerful communication system, one that is especially important early in life before language develops. An infant's cry of distress brings a mother running; a baby's beaming smile invites love and care."[10] Even after considerable language development, though, emotions are no less important for healthy, well-adjusted individuals. In fact, many of our most important language skills are related to emotional functioning—such as the ability to name what we are feeling and communicate that to others, or the ability to understand the emotional expressions of others and communicate that understanding in words or actions.

We punctuate our most important messages with displays of emotion, and emotions carry the most important aspect of messages that have both emotional and verbal content. If someone says, "I'm so happy to see you," yet all the while is looking at your companion, the nonverbal behavior is the tip-off for what is most important to the speaker at the time. Generally the verbal content of a message and the emotional component are consistent with each other, but not always.

When there is a lack of consistency, called a lack of congruency, we often experience the message as confusing, but the emotional component of the message is more reliable than the verbal content relative to what is uppermost in the mind of the sender. If I say, "Don't do that," while I am laughing, the laughter is interpreted as a more reliable indicator of my wishes than the words. The emotional aspect, in effect, countermands my words. Without emotions, some of our most critical communications would fail to influence others.

Within interpersonal relationships, emotions serve as the primary glue that holds relationships together. Feelings of love help maintain the most important human connections, but even feelings of fear contribute to solidarity, insofar as fear of losing close relationships can be extraordinarily motivating. Even hate may contribute to group cohesion by separating "our" group from "theirs." The role of emotion in the maintenance of important relationships cannot be over emphasized, so the importance of healthy emotional functioning cannot be over stated.

Intimacy and Empathy. Emotions are at the very heart of intimacy, and intimacy is the primary antidote for human loneliness. Intimacy is a process in which one person expresses an *emotional* communication; someone else receives the message, understands it, and then communicates in words or deeds that the message has been received and understood. In other words, intimacy occurs when an emotional communication is accurately received and responded to appropriately.

Understanding the emotional expressions of others is "empathy," and the ability to experience empathy is based, in part, on our own ability to experience a full range of emotions.[11] People who have learned to block out or ignore their own feelings are often unable to empathize with others. In turn, difficulty in labeling one's emotions, or the inability to express empathy for others, almost always has serious negative consequences. Violent acts of delinquency, for example, often are accompanied by an adolescent's inability to have compassion or understanding for others.[12] (Teaching empathy is now one of the hallmarks of many violence-prevention curriculums being taught in public schools.)

As humans, we all have a fundamental need for our emotional expressions to be understood. We need empathic contact with others, that is, we need to have our emotions acknowledged, and the reason is probably genetic or biological, an aspect of our "social nature."

From a therapeutic standpoint, an empathy-based relationship may be healing in and of itself. As a professional psychotherapist, I have noticed that many clients have never experienced empathic relationships in their homes, so a therapist's empathy often provides an excellent learning opportunity for these clients. Empathic responses help clients learn appropriate labels for their feelings, and empathic relationships can provide an opportunity for clients to learn alternative ways of expressing emotions for those who never learned appropriate labels or controls.

Because so many problematic behaviors are associated with traumatic events, there is a very close association between nondeclarative (emotional) memories and current behavior. Empathic listening, which focuses on client feelings, often provides one of the quickest strategies for identifying and understanding nonconsciously driven dynamics.[13] Among humans, and many animals as well, the most fundamental "truths" are those accompanied by strong feelings.

The ability to listen with an empathic ear can be a very powerful tool, but like all powerful techniques, there are potential dangers. First, within the psychotherapeutic community, I have observed clients who voluntarily remained in long-term therapy because of the highly reinforcing nature of an intimate relationship. Apart from any ethical concerns, overdependence on a therapist poses a risk because a client may not do the work necessary to find fulfilling relationships outside of therapy.

Earlier, we talked about reinforcers. Having our emotional expressions acknowledged may be one of the most powerful reinforcers known. However, people differ in terms of how they want their emotional communications to be acknowledged. There are many individual and cultural differences when it comes to the proper way of expressing emotion or acknowledging the emotions of others. I have met men who would rather die than cry. However, there are other ways of responding to emotional messages without relying on a verbal statement. Getting out of the way of someone who is angry is also a way of responding to an emotional communication. Thus, the conclusion still seems valid that we need to have our emotions acknowledged in some way.

Beyond always resorting to a verbal statement to express empathy, there is another pitfall to empathic communications, which involves how statements of empathy are made. As a supervisor, I have observed many therapists who inadvertently impeded therapy by how they phrased their responses to the emotional expressions of clients. Specifically, there is no such thing as a past or future emotion, by definition. Because emotions involve a physiological change, emotions are either happening now or not happening. Thus, when we respond with empathy to someone's expression of emotion, we can do so in a couple of ways.

Client Statement: I was so angry when my father yelled at me in front of my friends.
Undesirable Therapist Response: You *felt really humiliated* because your father belittled you in front of others.
More Therapeutic Response: As you talk about your father's outburst, I see the tears in your eyes, so it must still hurt.

Generally, the first response is less desirable. The key to understanding any emotional communication revolves around the emotion that is *currently* being expressed. Furthermore, if a goal of therapy is to teach a client about emotion, a client can more readily learn about his or her feelings when an emotion is occurring. "Talking about" a past or future feeling confuses the issue because the client is expressing a concept, not a feeling, or may be experiencing another emotion altogether.

Although various researchers do not agree on a standard definition of emotion, there appears to be agreement that emotional processes involve physiological changes within the body.[14] Because each emotion tells us something slightly different about our relationship to the environment, each emotion manifests its own pattern of physical changes within the body. Even the "same" emotion, anger for example, might show a different pattern of physical change from one moment to the next. Nevertheless, all emotions have in common that they produce changes in the body, in such ways as increasing or decreasing heart rate (depending on the emotion); changes in the chemical composition of body fluids like saliva and blood; changes in the skeletal muscles; faster or deeper respiration; and changes in blood pressure, pupil enlargement, and decreased gastrointestinal activity. No emotion triggers each response, but all emotions trigger *some* response.

When we talk about emotions, we are not necessarily experiencing the bodily changes that accompany them. Thus, if we are not having a physiological response, we are not experiencing the emotion. Clearly, talking about feelings can trigger a feeling, sometimes, but talking about a feeling is not the same thing as having one. Let me give a typical example of a therapeutic encounter.

A client is speaking about a past occurrence and keeps focusing on how happy the event was. The therapist might say, "As you talk about your dad's fiftieth birthday, *I hear you describe an even that was supposed to be festive, but you seem sad*." This statement illustrates what is called empathic mirroring. The statement communicates understanding of both content and feeling.

While on the topic of emotion and therapy, let me comment on what is considered a sacred cow among some theorists, that is, the idea of "therapeutic objectivity." As we come to understand more about the reasons why we, as humans, have emotions, the idea of emotional neutrality (objectivity) is based on a dated, fundamentally flawed theory of emotion, the idea that emotions invariably cause us to either lose control or lose "objectivity." On the contrary, properly understood, emotions provide us with an additional type of data that is not available from any other source.

Communications to Ourselves. Not only do emotional communications send powerful signals to others, they send signals to ourselves. When we behave in accordance with how we feel, the very act of behaving has a powerful influence on our thinking. Assume that a small child feels terrified at the sound of a tree branch scraping against the window at night. If the child resorts to hiding under the covers, the act of hiding accentuates the fear. Hiding tells the child, "Your fears are fully justified, you had better hide!"

It often happens that people do not recognize their feelings until they have behaved in a certain way. I have met clients and children who did not know what they were feeling, but because of their behaviors, their feelings could be reasonably inferred. We might act angrily and only know of our anger after we have yelled at a hapless store clerk. However, there is a potential pitfall here, as you might guess. People often infer the "rightness" of their actions on the strength of their emotions. However, the brain's emotional apparatus is not designed to rationally analyze external events, beyond signaling us that something *might* be important to our current needs.

Contribution to Personality. Beyond their vital functions for communication, emotions provide each of us with a distinctive mark of personality. The way each of us expresses, or withholds, our emotions becomes an enduring aspect of our individuality, and the ways we respond emotionally to others is an important aspect of being human. How we express our emotions and how we experience them also contribute heavily to our sense of self. Indeed, the experience of emotion is vital to the experience of feeling alive (which is not to say that some people are not troubled by their feelings, particularly feelings of sadness or loss).

When individuals do not experience emotions, or experience only a very restricted range, that restrictiveness can cause problems for the individual, just as easily as excessive emotionality. People who do not experience a full range of feelings often report chronic boredom.[15] Boredom, in effect, is the opposite of emotion, the absence of feeling. Each of us has experienced boredom at some point, and it is rarely described as a pleasurable experience.

Emotional Memories

Emotions appear to facilitate, or co-occur, during memory formation, such as fear conditioning. However, it is not likely that entire, full-blown emotional reactions are stored in memory. Because of an emotion's "action tendency," which relies on physiological changes, it is unlikely that an emotion can be stored, per se. Rather, the brain stores aspects of an emotional event in component parts, and afterward the emotion can be retriggered. Over time, the exact same physiological pattern may or may not be retriggered every time the "memory" is triggered. Clearly, some highly emotional events lose strength after enough time has elapsed. Immediately after the death of a loved one, we are usually disconsolate; but years later, we may still remember the person and the loss, but with far less intensity. On the other hand, even after the emotion has faded, the nondeclarative memory of the event may still influence our behavior. We might have lost someone in a tragic car accident and, thereafter, avoid the intersection where the accident occurred, long after the profound grief has dissipated.

Emotional memories, as a topic, are difficult to sort out because of the controversy surrounding these types of memories. However, it is clear that when people are subjected to chronic stress or trauma, especially during times when the brain is developing, brain physiology can be influenced, in which case, individuals so affected may forever be more reactive to emotional triggers.[16] Thus, we might argue that "reactivity," as a result of learning, is a type of memory, but it is more akin to the type of developmental changes we discussed earlier, and not a stored "emotional" memory, per se.

Emotions and Mood

In everyday usage, we are most likely to talk about our feelings. Feelings, in turn, are typically used as synonyms for "emotions," at least in everyday usage.

When people talk about their feelings (or emotions), they often refer to "mood" as well. "I don't feel like it." "I'm not in the mood." "Don't bother Dad, he's in a bad mood." "Mother's mad, so leave her alone." Each of these statements expresses an idea about how someone is "feeling," and these statements imply that mood and feeling (and presumably emotion) are viewed as interchangeable entities.

So, are feelings, emotions, and mood the same thing? No. Feelings and emotions can be used synonymously (at least in the following discussion), but there are important differences between emotion and mood.

Emotions are relatively brief.[17] Our emotions come into existence for a specific purpose, to inform us, to signal others, to get us ready for action, or to energize our behavior. Moods may last for several days, weeks, or even months, and they rarely come into being for a purpose that we can identify. In our day-to-day existence, we use the terms "mood" and "emotion" (or feeling) synonymously, which leads to a misunderstanding of both emotions and mood. Although we might say, "Dad's in a bad mood," what is meant by that is not always clear outside of context.

If we are being sticklers for definition, Dad could be in a bad mood, if he were depressed or anxious, but if he is just a little out of sorts today, or this morning, it would be more accurate to say, "Dad is *feeling* a little irritable this morning." On a day-in, day-out basis, we rarely make this distinction. Generally, among professionals who treat problems of depression, the two most common moods for which people seek treatment are depressed mood and anxious mood.

Moods also exert negative influences on our emotions. If I am depressed (a mood), my mood will decrease the likelihood that I will experience joy or happiness. That is unfortunate because emotions, like joy and happiness, serve important functions, so the loss of a whole class of emotions (the happy ones) is a serious loss to my perceptual and analytical abilities, not to mention the loss of the simple joy of living. In a related matter, depression increases the number of painful emotions we are likely to remember or experience. That is, we are more likely to remember sad events during a bout of depression than remember past joys.[18] The term for this type of memory bias is called "mood-congruent" memory. In short, moods not only restrict our range of emotions, but they influence what we will recall, which also colors how we view the world.

Another difference between mood and emotion is that emotions have relatively specific facial, physiological, and muscular characteristics (depending, of course, on the emotion). When we are happy, it shows on our faces. Moods, on the other hand, do not have closely defined behavioral and physical features. Although both anxious and depressed moods alter physiological functioning to some extent, moods do not show overt facial and behavioral reactions like emotions.[19] Clearly, depression may manifest as lethargy, low energy, or irritability, but it does not have a "typical" facial characteristic, whereas emotions often do. (This is a general rule, though, and there are exceptions.)

Emotions are always associated with specific events or objects. When we experience emotion, we do so in response to a specific cause. We become sad at a loss, happy at a find, angry at a slight, fearful at a threat, and so on. Moods, in contrast, rarely correspond to a single precipitating event, with the possible exception of a catastrophic loss, like the death of a loved one, for instance.

Another difference between mood and emotion pertains to our bodies' reactions. Emotions typically result in a "readiness to act."[20] Moods rarely prepare us for anything (except maybe to sleep). Depressed individuals often cannot act, except with extreme difficulty. Anxious people (those with an anxious mood) often "act" through avoidance.

In general, as noted, when we are depressed, we are more likely to see the negative in everything. When we are anxious, we are more likely to see a mugger hiding in every shadow. Moods distort our perceptions and decrease the likelihood that appropriate actions will be taken. Emotions, in contrast, add to our perceptions and enhance our thinking. Emotions are experienced in response to an event and provide us with information about our relationship to the world around us. If we experience fear at the sight of a moving object, our fear is a valuable warning.

Complex and varied emotional functioning is one of the hallmarks of healthy human functioning. The ability to feel and express a full range of emotions is one of the most fundamental criterion for defining mental health.

BRAIN STUDIES

By any standard, emotional processes are complex, requiring a coordinated response that usually involves the brainstem, limbic system, and neocortex (and at least one expert adds the hippocampus, another structure within the limbic system).[21]

The Neocortex

The neocortex, described previously, is thought to play a key role in emotional processing, by integrating diverse sources of information into a full emotional response, and also by helping us determine how to respond.

For many years, the right hemisphere of the brain had been thought to be the "emotional side" of the brain. However, more recent studies have shown that the frontal regions of both hemispheres participate in under-standing the emotions of others and expressing our own emotions. How-ever, researchers have also learned that "negative" emotions are processed mostly in the right hemisphere, and "positive" emotions are largely pro-cessed in the left hemisphere.[22]

Limbic System

The limbic system evolved long before conscious "reasoning powers," and some authors believe that the limbic system evaluates the world like a separate brain and forwards that information, which is limited to an emotional reaction, to the neocortex for further analysis and integration. It may even be possible for us to have an emotional reaction to an event, apart from our conscious knowledge of the event. Many experts believe this is the case.[23] However, this point is highly controversial, and there is nothing close to a consensus on this point among the experts. It is probably safest to conclude that aspects of an emotional reaction *might* influence behavior outside of awareness, but whether we can have a full-blown emotional reaction, including the concomitant physiological changes, and remain unaware is debatable. (Nevertheless, the assumption that emotions *can* influence people outside of their awareness *is* a common belief among many psychotherapists.)

The limbic system has long been considered the seat of emotion, but that distinction is somewhat misleading. As we have seen, it takes much of the brain to produce behavior and mental processes, and the mammoth job of coordinating complex, emotionally driven behaviors certainly requires many parts of the brain working together. Nevertheless, the limbic system continues to receive a major share of the research attention relative to emotional functioning, and within the limbic system, the amygdala is the star of the show.[24]

Amygdala

The amygdala is an almond-shaped cluster of nuclei that is believed to produce the experience of fear, and probably anxiety as well.[25] By default, the amygdala is also one of the likely candidates for other feelings as well, but the experience of fear is the most well-documented connection. How-ever, from a theoretical standpoint, the amygdala is described as having the necessary neural connections to the neocortex and other brain areas to make it the top candidate for contributing to emotional reactions.[26] This view is not without its critics, however, so it is probably safest to conclude that the exact cause of our "feelings" is still a matter of debate, but when we learn to associate some event with a feeling, the amygdala is almost certainly involved, and the experience of "feelings" may also be dependent on the various nuclei that collectively constitute the amygdala.

The amygdala is also part of our body's "first-alert system." When we are alerted and become oriented to environmental events that have personal significance for us, the amygdala is likely to be involved.[27]

Because the limbic system is associated with the experience of emotion, the system is often depicted as one that only receives general information, awaiting a more refined analysis from the neocortex. However, the limbic system receives information that is already highly refined and reacts to that. For example, the amygdala will respond to a familiar face, but it will not respond to a face that is not familiar. The ability to recognize a face stems from the neocortex, but since the amygdala responds to a known face, its role in emotional processing may be that of a relay station, to provide a link between bodily sensations and information that has already been processed by the neocortex.[28]

In sum, although many researchers generally agree that emotional experiences involve the amygdala to a greater or lesser extent, they agree on little else. Like all cognitive functions, no two emotional reactions draw upon identical neural pathways, so other parts of the brain have at least supporting roles in the production of emotion. The degree to which those other structures are involved is not known, at present, but the matter is certainly debated widely.

The Brainstem

The brainstem is mentioned briefly, in passing, because it contributes to many of the bodily sensations that we experience as emotion, like the sinking feeling we get at first realization of a loss.[29]

The foregoing discussion of the brain's role in emotional functioning has been quite brief, partly because the exact nature of the brain's role is still a matter of contentious debate among the experts. When it comes to highly complex emotional states, like "love," or "pride," next to nothing is known about the neurobiology that underscores these feelings.[30]

Now, we are ready to consider "The Emotion Principle."

PRINCIPLE 11: THE EMOTION PRINCIPLE

Our emotions, collectively, serve many vital functions. First, they provide us with an innate means of communicating our deepest needs, from birth to death. Second, our emotions provide us with an indispensable means for "labeling" which events have personal significance. Third, our emotions prepare us for action, and when action is necessary, emotions provide us with drive and energy. Finally, our emotions reinforce some of our actions and punish others.

Emotions, collectively, serve many functions that are not provided by other brain mechanisms. As we would expect, they draw upon many regions of the brain, cutting across large sections of the brain, to include the brainstem, limbic system, and neocortex, not to mention all the sensory inputs that first trigger a recognition of an emotional event. It is not surprising that emotions provide so many different functions because they appear to activate such large portions of the brain.

Historically, emotions have received relatively little scientific attention, but there have been notable exceptions. Recently, there has been a tremendous increase of interest in emotional processes, and research on the topic is proliferating. Possibly as a result of that research, we may discover that the phenomenon we have tagged with a single label, "emotion," is really many different phenomena. One thing we have learned, though, is that our labeling system for many brain functions has grouped diverse functions together with a single label, like "emotions," so it would not be surprising if emotions were eventually described as several diverse processes.

IMPLICATIONS FOR UNDERSTANDING BEHAVIOR

Emotions during Prehistory

It is interesting to speculate about the functions that emotions served for our prehistoric ancestors. We can make a few inferences from the fossil evidence, at least from the time when the human skull had reached its current size. Since emotions serve so many vital functions for us today, it is likely that emotions served the same or similar functions among our prehistoric relatives. Certainly, we can see the advantage of having a system that would quickly orient our ancestors to important environmental events and energize behavior when action was needed. We can also see the advantage of a communication system that reached maturity before the advent of language.

Experts do not know when human language first emerged. There is no doubt that language abilities were possible from the time when the first genetically modern humans emerged, and it was *probably* possible for *Homo erectus*. However, it is now known when language first appeared, so it is likely that emotions served a vital communication function among prehistoric humans before language abilities emerged. In short, emotional expressions were probably very important communication tools for our prehistoric relatives, probably every bit as important as emotions are for us today.

Emotions Motivate Behavior

Emotions motivate many behaviors, both directly and indirectly. The direct ways are most obvious. If a little boy hits his younger brother during a fit of rage, his behavior was motivated by anger. However, even thoughts about future emotional consequences can influence our behavior. The desire to avoid a panic attack, for example, can motivate someone for a lifetime, even when panic is not currently being felt.

As discussed earlier, if a behavior is motivated by a need, such as the need for emotional understanding, the behavior may become highly resistant to change. Emotionally driven behaviors are powerful motivators. These behaviors can persist until the underlying need is satisfied, or *until* another behavior is substituted that more effectively satisfies the need.

Substituting one behavior for another that meets the same need is one of the soundest principles underlying behavior change. When we learn to clarify our needs, we can also learn more effective ways to meet our needs, and the best means of clarifying our needs is to understand our own emotions. When therapy is based on these principles, resulting changes can be quite effective.

Because emotions motivate our behavior, we can use that energy and knowledge to make changes in behavior or a situation. Perhaps, like me, you have ignored an annoying habit of a close friend, only to be driven to a point where you could not stand it any longer? "Stop that!!!!" These words, expressed emotionally, are energized by emotion. The natural tendency to keep quiet was overcome by the strength of the emotion.

Some emotions reward behavior. If we take food to a shut-in, the feelings we experience may be quite positive as the grateful individual thanks us for our help. Our behavior has been reinforced by the positive feelings that ensue. Therefore, the behavior is more likely to occur again. If we yell at a motorist in traffic, the release of tension is reinforcing. Thus, our yelling is more likely to occur again.

Because emotions involve physiological changes in the body, they are powerful drugs. Emotions supply each of us with a large internal pharmacy that can be employed to reward behavior. Although we cannot consciously create our emotions, we can intentionally engage in behaviors that will produce specific *feelings.*

The knowledge that certain behaviors will result in predictable feelings is a powerful tool that we can employ to make changes in our lives. This principle is often relied upon to treat depression and anxiety. In the case of depression, for example, a client and therapist can make a list of activities that have a high likelihood of producing pleasure. If the client cannot experience pleasure, then a list of activities can be constructed that will distract the person from morbid or sad thoughts. When the client engages in those behaviors, the depression may weaken. This technique is quite effective.

By way of warning, I should mention that there are individuals who have become self-described "adrenaline junkies." It appears as if "thrill-seeking" behavior is often motivated by the need for "another fix." Thus, there is a downside to using our emotions as a means of reinforcing behavior. Because our emotions are so powerful, like any other reinforcers, emotions can be abused.

Emotions Punish Behavior

Several emotions appear to exist to punish behavior. Guilt, shame, fear, and anxiety are all examples. Each of these emotions can be generated by our brains in response to behavior. If we hit the dog out of anger, we may experience guilt afterward. Guilt, in turn, may inhibit us from doing the same thing again.

Research has consistently found that negative emotions, defined here as anger, fear, sadness, and disgust, often take precedence over positive emotions and over other events as well.[31] Earlier we learned about the brain's propensity to attend more readily to the negative, which was called automatic vigilance, and by the same token, negative emotions also overshadow other mental processes.

Even if people are just weighing future possibilities, negative outcomes are given more weight than positive gains. Even a single negative feeling about someone may overshadow our ability to see positive characteristics at the moment we are experiencing the negative feeling. (This finding, I should add, coincides exactly with what we learned earlier about negative stereotypes.)

Although I have heard people categorically state that they would like to eliminate unpleasant emotions, each "negative emotion" has an important place in healthy human functioning. Without fear or anxiety, many of us might not be alive today. This conclusion is based, in part, on the knowledge that fear and anxiety override positive emotions. They do so because they are controlled by different brain pathways than other emotions. So-called negative emotions presumably survived millions of years precisely because they provided survival advantages that were not produced by other emotions or brain processes.

If a prehistoric ancestor was out hunting, and if he was too overjoyed by the sight of a rabbit that he did not heed the growl of a hungry wolf, he would not have lasted long. Fortunately, in this case, the sound of the wolf's growl would immediately take precedence over the joy of having found a rabbit.

Without question, negative emotions can become so chronic that they exert a deleterious effect on behavior. The example we are most familiar with is that of people who are hostile or chronically angry. It may be that hostility is really more akin to a chronic mood, but it is certainly true that some people react in anger frequently, with negative consequences. Just as a negative mood, like depression, tends to distort perception, some emotions can also become so common that they, too, may fail to perform the signaling functions for which they presumably evolved.

Modification of Our Environment

Whenever we have an emotional reaction, we are often interacting with the world around us. In fact, interacting with the environment is considered a primary reason for the existence of emotions. Just as the environment impacts us, our behavior can impact the environment. If we suddenly jump at the rapid approach of a cockroach, our revulsion can be overcome by a quick rush of anger. By smashing the bug, we have altered the original stimulus considerably! I raise this issue here because one result of emotions is their ability to put drive into our behavior, allowing us to do things that we might not do otherwise.

Changing and Redirecting Emotional Energy

There are many ways to impact our emotional functioning, and you will be able to add to the following list when you get the hang of it. First, I will recap what emotions are, because knowledge about them is the key to using them for change. Emotions are complex reactions to the world around us, and emotions involve the following components: (1) there is an environmental event; (2) there is our brain's preliminary assessment of the event; (3) if the situation appears to have potential relevance for us, there is a physiological change within our bodies that we experience as a feeling; (4) there is our bodies' first behavioral reaction to our brains' initial assessment, which may include "action tendencies"; (5) there is our brains' further and continued evaluation, which adds to our ongoing behaviors; (6) there is the impact of our behavior on the environment; and (7) there is our brains' continuous reevaluation (which includes past learning about similar events).

An emotional reaction involves the whole complex summarized earlier. If we change one single component—any component—we can substantially alter the whole complex. (By way of warning, though, before any of us begin to change our emotions for any reason, we should be certain we understand the reason[s] for our feelings, because one of the primary reasons for our emotions is to signal us.)

As we can see, some components in this model are more amenable to our direct control than others. However, with our new understanding, *if* we want to modify how we are feeling, or wish to modify our emotional behavior, we can do so in a number of ways.

By looking at each component of an emotional reaction, especially those over which we can exercise some control, we can consider what might happen if we altered that component.

Changing the Environment

If we alter a precipitating event, we can create an emotion or modify an emotional behavior. In fact, we change our environments all the time with the intent of impacting how we feel. If you have ever come home from an exhausting day at the office and turned on your favorite music, you have changed your surroundings with the specific intent of altering how you feel.

Once an environmental event has triggered an emotion, our emotional behavior can impact the original event. Altering the world in which we live is a major outcome of emotional behavior. If you stub your toe on the living room chair and angrily push the chair out of the way, you are using your anger to modify the environment and perhaps avoid a future injury. Changing things around to suit ourselves is one of the easiest ways for us to impact our emotional behavior.

Altering the Feeling

Altering feelings is one of the most common ways people *try* to impact emotional functioning. This tactic may work for some, but I rarely recommend it. Feelings are seldom under our conscious control. Furthermore, attempts to weaken or ignore our feelings can actually *increase* their strength. If our emotions are intended to signal us, then by ignoring them, we would expect an increase in their strength.

People with severe anxiety, for example, often recount attempts to *control* their anxiety, with the result that they often become *more* anxious. (Earlier we discussed the technique of "flooding," in which individuals "face their fears," until the fear subsides. This technique often works for changing a feeling, but it does not involve an attempt, per se, to either ignore a feeling or weaken it. Rather, the individual is encouraged to walk headlong into the fear.)

Changing Your Reaction

Although our bodies prepare us to act, largely automatically, we can change our initial reaction to an emotionally arousing event. Changing our initial response will quickly change how we feel and will also affect the communications we send to others. By changing our first response, we may also alter the precipitating environmental event (discussed above). Finally, by changing how we initially respond to an event, we often influence our thinking.

When it comes to our behavior, we have so many options that it is not possible to enumerate them all, but I have provided several examples here to help us begin to think about the possibilities.

First, remember the rule of thumb: If we alter a characteristic behavior, we often alter an emotion. If we respond differently from normal, we will often change how we are feeling. In turn, our emotional behaviors will change. It really does not matter whether we alter our first reaction or our ongoing behavior. (If a new behavior *does not* change how we feel, we should perhaps try sticking with the new behavior longer. Sooner or later, the feeling will change and the whole interaction will be impacted.)

How we behave alters how we see a situation. A well known principle often touted in self-help groups is, "Fake it until you make it." In psychological jargon, act "as if." If we behave in a manner consistent with how *we* *want* to feel, we will come to feel that way. If we are terrified, but act fearlessly, we can come to feel fearless. This strategy may seem ridiculous, until we consider how common this strategy really is. None of us start out as experts in anything. Rather, we always "fake it" at first. Even accomplished surgeons have to make their first incisions without help. We all begin as novices. Only practice catapults each of us into the ranks of experts. So, we can use our emotional energy to cross whatever barriers might be holding us back.

Another helpful way to alter a first response is by learning another response, and then substituting it for an old one. This technique works amazingly well, and is the basis of many therapeutic interventions.

If we suffer from chronic fear, we can take a self-defense or assertiveness class. Once we have learned to react initially with a strong behavior, subsequent action may not be necessary.

If we feel really angry, anxious, or fearful, we can alter our emotional reaction by employing a well-rehearsed relaxation response at the first moment of anxiety. Relaxation can compete with some emotions. For example, it is not possible to feel highly anxious and fully relaxed at the same time. Similarly, it is not possible to be extremely angry and fully relaxed simultaneously. (We employ the term "ambivalence" to imply that we have "mixed feelings." However, it is unlikely that we actually have two different feelings at the same time. Rather, our feelings may fluctuate, or we may have conflicting thoughts.)

Another technique for altering our first response during an emotional interaction is called a "time-out." This technique involves removing a child from a provocative situation or channeling our energy into a more constructive action. Time-outs are usually recommended for individuals who have problems with temper or violence, but the principle can be learned and employed by anyone. Typically, a time-out involves a substitute behavior, like taking a walk (in lieu of some other, more harmful behavior). We can practice the technique before we need to use it, much like learning to swim before we fall into deep water. Then, when we encounter a provocative situation, we simply substitute our new behavior.

One of the most valuable ways to alter emotional behavior is to learn to talk about our feelings. Talking is a behavior, one of the most important behaviors we can employ to our advantage. When we talk about feelings, a number of positive results can be achieved. First, talking about an emotion often dissipates or weakens the emotion's energy. If we are filled with tension or anger, telling somehow how we feel can alter the feeling. By talking about feelings, we may actually shift the locus of control within our brains from a limbic-centered response to one centered in the neocortex. (At least, this is one theory.)

Clearly, though, talking about feelings is not the same as feeling. When we talk about our feelings, we make the process more intellectual. The process becomes less emotional, and the intensity of our feelings may dissipate.

Talking about feelings also has an added benefit. Telling others about our feelings is a way to achieve a sense of closeness or intimacy with them. When we share our emotions, even unpleasant ones, others often report feeling closer to us. (I should mention, there are many exceptions to this generality, and some people cannot tolerate close emotional sharing.)

Talking about unpleasant feelings does not come easily for many of us. The "fear of talking" is probably rooted in negative past experiences when your feelings were not acknowledged by others, or worse, when your feelings were ridiculed. Also, there are vast cultural and individual differences relative to the degree to which emotional expressiveness is considered appropriate. Thus, many children are taught to withhold displays of emotion, at least under certain circumstances.

For those of us who are reluctant to share our feelings, we can find "safe" people in whom we can place our trust. Sometimes, self-help groups are useful because most of them have strict rules intended to produce a safe atmosphere in which people can share safely what they are feeling. If self-help groups are unavailable, we can seek the services of professionals or the clergy.

Alter Your Thinking

By altering our thinking, we substantially impact our emotional functioning. Although many people are aware of this fact, few of us learn to control our thought processes to the extent possible. In fact, one of the biggest contributions to human suffering may be the failure of people to maximize their thinking abilities. Our emotional processes almost always work flawlessly, but our thinking processes often interfere with optimal emotional functioning.

Since all emotions involve a physiological component, many feelings share a similar state of physiological arousal. Often, the biggest difference between two emotions is how we label our feelings. In fact, people often rely on situational cues to determine how they are feeling. Two emotions that are "first cousins" to each other are fear and excitement. Thus, if we feel slightly fearful or anxious, we can choose to think of our feelings as excitement. (Normally, I do not recommend that you mislabel any feeling, but this is one exception.) By relabeling mild fear as excitement, we can become excited. Once excited, we can accomplish a task that otherwise might have been avoided.

If we are fearful about an upcoming speech, we can tell ourselves, "This is really going to be exciting." Focus on the excitement, on the adventure. We can learn to enjoy our feelings, and taking even a small sting out of fear can have a positive influence on behavior.

Sometimes at the gym, I put on earphones and listen to inspiring music. I then imagine myself at the head of a pack of runners, moments away from clinching a gold medal. I feel a surge of energy course through my body, and I use that energy to spur me on to greater levels of accomplishment. With practice, we can get good at this sort of "emotional" manipulation to reach goals. This is a form of visualization that can even be used therapeutically.

Handling Other's Emotional Behaviors

What is the best way to deal with others' emotional behaviors? The answer can be given in one word: acknowledgment. One of the biggest mistakes we can make when responding to the emotional outbursts of others is to assume that acknowledgment and agreement are one and the same, but they are not.

We can learn to acknowledge someone's emotional message without agreeing with the message. In fact, this is among the most important of all communication skills, in my professional opinion. If someone is yelling, "You're a crook, you stole my money!," our first impulse is to disagree with the person. "No, I'm not a thief!" However, we must look at what drives all emotion: Emotions arise when an event is important to an individual. Whether people are right or wrong, the point is that an event has personal significance for the individual who is upset. Thus, we can acknowledge an emotional message without agreeing with the accuracy of the words. "I can see you are outraged at me. Tell me what happened." This response does not support the individual's accusation, but it clearly acknowledges the emotional communication.

Emotions and Defense Mechanisms

The failure to consciously register emotion has often been attributed to "defense mechanisms." However, if emotions can indeed fail to register on our consciousness, there is an alternative explanation that can be understood by a simple conditioning model. When the emotional displays of infants are chronically ignored or misjudged, those emotional displays become blunted. Over time, an infant's passive response style may become permanent. As adults, these individuals may be totally unaware of their own feelings because their emotions were never mirrored appropriately— their emotional displays were not reinforced.

In many dysfunctional families, children receive payback only by learning to heed the emotional displays of others. If one's father often came home in a rage, there would have been no survival value in knowing one's own feelings. It would have been much better to always know how Dad was feeling.

In many families, children's displays of emotion are harshly punished. When a child's emotions are chronically punished or ignored, it is only a matter of time before all outward signs of emotion are extinguished. If that happens, an individual may no longer pay attention to internal bodily sensations or make the connection between those feelings and a specific emotion. The connection between an internal bodily state, like fear, and the word "fear," must be learned. Through a simple conditioning model, we can see how emotions can lose their ability to function optimally. If we do not pay attention to them, if we do not have labels for them, or if we fail to appreciate the emotions of others, then a powerful communication tool has been blunted.

CONCLUSION

Emotions serve many vital functions for humans. They provide us with an innate means of communicating our deepest needs and desires. They also serve to distinguish one individual from another. How each of us learns to express or withhold our emotions becomes an enduring hallmark of our individuality. Emotions provide us with energy. Moreover, emotions orient us to important events in the world; they punish or reinforce our behavior; and they share an intimate relationship with certain types of learning and memory.

Emotions are absolutely vital for normal and healthy human functioning, and their diverse characteristics are embodied in "The Emotion Principle," restated here.

PRINCIPLE 11: THE EMOTION PRINCIPLE

Our emotions, collectively, serve many vital functions. First, they provide us with an innate means of communicating our deepest needs, from birth to death. Second, our emotions provide us with an indispensable means for "labeling" which events have personal significance. Third, our emotions prepare us for action, and when action is necessary, emotions provide us with drive and energy. Finally, our emotions reinforce some of our actions and punish others.

* * *

Our emotions provide us with indispensable information about our relationship to the world. Our emotions also provide a valuable type of information to our neocortex, which must integrate sights, sounds, smells, and emotional reactions and then formulate a plan. By devising a plan of action, the neocortex helps to moderate emotional reactions by providing a range of options. We can "think about" a situation and decide what to do. From this perspective, as part of a complex reaction, our emotional responses contribute to our powers of thought, and it is to the topic of thought that we will turn now.

CHAPTER 10   TIME TRAVEL

Have you ever dreamt about taking a little peek at *next month's* stock market report? That would be a handy little trick, wouldn't it? Or perhaps you have wanted to go back when times were less complicated, maybe take a horseback ride through the western United States during the early 1800s. The thought of time travel is one of those science fiction ideas that triggers just about everyone's imagination. But is it really science fiction?

Some people go into the future all the time. Stock analysts, weather forecasters, astronomers, scientists, and the list continues—all working in the future to some extent—and the best do so with amazing results. Any of us can go into a bookstore today, buy a calendar, and know when the next full moon will occur, or the first full moon in the year 2050. People engaged in manufacturing can plan ahead one month and know with a high degree of confidence what will roll off the assembly line in thirty days.

There are others, like myself, who go backward in time. Historians, biologists, detectives, anthropologists, psychologists—all people who help individuals understand the present by reconstructing the past. (In places like Santa Fe or Sedona, I am told, there are even those who visit past lives.)

We humans routinely travel through time, and as far as anyone knows, we are the only species that can contemplate different future possibilities or dwell on the past through a unique collection of abilities called "thinking."

Just a few years ago, the ability to worry about the future or dwell on past misfortunes was considered the hallmark of neurosis. That label may apply to the problems of some, but on balance, the ability to live outside the present has provided humans with an *unparalleled advantage* in the game of survival. This advantage, however, has not been achieved without some nasty side effects along the way, over and above the problems of "neurosis."

World history is replete with examples of madmen who imagined a different world and proceeded to carry out their vision through programs of mass destruction. Fortunately, others have used their vision to make the world a better place.

From an evolutionary perspective, human reasoning abilities are new indeed. In fact, of all the brain's capabilities, our ability to think is new enough that we can roughly estimate when it first appeared. We will review what we have learned about our past and then try to pin down the birthdate of this extraordinary gift.

THE EVOLUTION OF THINKING

One of the conclusions we reached about our past is that the brain reached its current size *before* the phenomenal changes began to occur that are associated with human beings—innovations like art, literature, agriculture, animal domestication, airplanes, and so on. We also learned that there was *not* an immediate increase of inventiveness when the brain's large size first appeared. In fact, there appears to have been a pretty big lag time. The acceleration of human inventions did not show a dramatic rise until after permanent settlements began to spring up, sometime between ten and fourteen thousand years ago. This information, though, creates an interesting riddle. Since the brain reached its current size, perhaps two hundred thousand years ago, why did it take humans so long to "modernize," to start changing things around to suit them? We may never know, but several ideas are relevant.

Although anthropologists can pin down brain size quite accurately from skull fossils, they cannot unequivocally assert that larger brains instantly endowed their owners with the intellectual capacities we have today. It may have taken a long time for human intellectual capacities to fully develop, even after the brain reached its current size. Assuming that learning is one of the mechanisms that contributes to innovation and intelligence, it may have taken thousands of years for the first big-brained humans to learn enough on their own without the advantages of skilled teachers to really challenge newly developing intellectual capabilities. Further, given the weight of tradition, people may have felt compelled to follow the "old ways," in spite of the ability to innovate. Finally, we should not discount the very real possibility that prehistoric bands *liked* their lifestyle and wanted to preserve it.

Perhaps it is ironic, but other capacities already arising from within the brain may have thwarted innovation (at least at first). We cannot assume that new inventions or ideas were automatically seen as positive, no matter how clever those ideas may now appear in retrospect. Some individuals may have been quite inventive, but their inventions may not have taken root. To understand why, we can try a little experiment. Get your family to move a traditional holiday dinner from where it has always been held, and see how easy it is to mess with tradition. Convincing any group to alter a long-standing tradition is often difficult or impossible.

Without question, though, there were some significant innovations during the period of human brain growth, like improvements of stone tools, but it cannot be assumed that innovations were instantly adopted. When the momentum of custom and fear of change are added together, a potentially powerful force for maintaining the "old ways" is in effect.

Over time, though, given humans' incomparable reasoning abilities, some innovations were inevitable. Also, because of the brain's almost unlimited ability to copy and learn from others, once an invention did prove its superiority, its renown would have spread rather quickly. Perhaps there is irony here—the brain's innate fear mechanisms would have worked *against* change while simultaneously *compelling* change! On the one hand, anything new would have been feared, and therefore shunned. On the other hand, when a new technology appeared to improve survivability, a band might have been driven by fear alone to adopt the new technology.

The brain first reached its current size within hunting and gathering societies. In the broader scheme of things, the brain has lived most of its existence within hunting and gathering societies. If we look at the modern brain within a modern body, and exclude *Homo erectus*, the brain has still spent an estimated one hundred thousand to two hundred thousand years engaged in hunting and gathering, and only ten thousand to fourteen thousand years in other pursuits. Thus, given the brain's comparative longevity within hunting and gathering societies, it is reasonable to conclude that it was well adapted for that lifestyle. If the foregoing conclusion is true, why did humans abandon hunting and gathering?

According to anthropologists, environmental changes, including climate changes (and possibly human population growth), forced humans to change. The Ice Age was ending. Seas were rising. Many traditional food sources had gone extinct. The rising seas changed the face of the earth. Changes from hunting and gathering to agriculture were necessary, or at least a new lifestyle was easier and more adaptive.

Regardless of why humans experienced a change from hunting and gathering to agriculture, the experts agree that the domestication of crops and animals and the adoption of a more pastoral lifestyle created major social changes. The ability to grow food and store it resulted in permanent settlements. Stable food sources and permanent communities resulted in larger populations. Larger populations then resulted in other changes. One was the uneven distribution of wealth, which meant that some individuals had more time to devote to greater thinking pursuits, which perhaps increased inventiveness.

Although some inventions, like better tools, *predated* widespread settlement, it was *after* the rise of agriculture and the rise of permanent villages and cities that humans started to show the range of their creativity, to include such things as the use of metals, the development of the first large cities, the invention of written records, the creation of widespread trading practices, and so on.

In sum, it seems likely that the brain was *capable* of sophisticated reasoning, perhaps for thousands of years, but climate changes (and possibly population growth) served as the indirect impetus toward increased human inventiveness.

Although the brain's physical growth may have ended perhaps as many as two hundred thousand years ago, the brain did not *lose* many of its earlier capabilities (as long as those capabilities continued to serve adaptive functions). This is a recurring theme because it is a fundamental brain principle. This is relevant here because the brain supports *many* thinking systems. Once the neocortex was added to preexisting brain structures, the brain's increased mental capacities began to show tremendous diversity.

When "high level" reasoning and thinking capacities are the focus of discussion, clear distinctions become blurred, and "thinking" and "learning" become almost indistinguishable from each other. I have spent many hours thinking about this chapter, yet I will not forget my thoughts when I stop for lunch or rest for the night. This must mean that some of my "thinking" has also been stored in memory. Storage, in turn, also results in "learning." Of necessity, therefore, this chapter will contain comments that pertain to both thinking and some forms of learning that we have not covered yet.

Researchers have speculated and written about the brain's many cognitive (knowledge) systems and have devised some general labels to describe them. Some of the issues are not fully resolved at present, but a discussion of those issues will help us think about "thinking" in a new light.

INSTINCTUAL BEHAVIOR

The concept of instincts may seem misplaced in a discussion of thinking, but many behaviors are attributed to instincts that are not instinctual at all—behaviors that are perhaps maintained through learning and thinking processes, so a theory of human instincts must be addressed.

The following anecdote of a nesting goose describes what an "instinct" looks like in action. The quote is apropos to us because humans are constantly depicted as having instincts. "The animal then extends its neck to fix its eyes on the egg, rises and rolls the egg back into the nest gently with its bill."[1]

This quotation describes a mother goose when her egg is accidentally pushed from the nest. She looks around, spies the errant egg, then gets up, walks around to the other side of the egg, then herds it back into the nest with her bill. She appears selfless, devoted, and vigilant.

None of us have any problem recognizing the goose's behavior as the "mothering instinct," but there is additional information in this study. The "egg" in the example was, on some occasions, a beer bottle! A nesting goose will rescue any round object that rolls from its nest, which illustrates one key point about instinctual behavior: It is rigid and *stereotypical*. All members or same-sexed members of the species perform an instinctual behavior in the same way every time a certain provocation occurs, like the "egg" rolling from the nest.

Instinctual behavior unfolds like a row of dominoes that have been stood on end. Once the first one is pushed over, the rest will topple. A process has been set in motion that runs its course.

In spite of the common and widespread belief that humans have instincts, there are few, if any, examples within the scientific literature that illustrate a behavior that unfolds in precisely the same way among all members of the human species, with some exceptions.

It is true that newborn infants, for example, often show behaviors that must be inborn. They respond to human and nonhuman objects very differently. If a nonhuman object is presented to an infant, the neonate will look at it while remaining alert. If the sound of a human voice is detected by the baby, the neonate will search for the sound and will react differently than it would toward an object. Toward a human, the infant's eyes will become wider, and the baby may stretch out her neck and lift her chin toward the source of a voice.[2]

At present, it is probably safest to conclude that humans may have a *limited* number of instinctual behaviors, but as we grow older, instincts become much less of a force in underlying behavior. This is a controversial issue, and clearly, instincts are commonly attributed to adults. However, if there was actually a "survival instinct," then all humans lost in a snow storm would innately know how to build a shelter, how to acquire food, how to generate warmth. The human desire to survive is a drive, not an instinct. The difference between a drive and an instinct is that a drive is a biological imperative, like gasping for air after being submerged in water for forty seconds. An instinct, in contrast, provides an individual with an actual *pattern of behavior*, like when the goose turned its head, got up, and pushed the "egg" back into the nest.

If there was an actual human "mothering instinct," to cite another common example, we would see something akin to universal mothering behavior, but there are no such examples among humans. One reason I belabor this point is that many human behaviors are attributed to instincts, and even professionals use "instincts" to explain behavior, in the absence of supporting data for the types of instincts being hypothesized. If we assumed that "mothering" unfolded naturally, instinctually, the assumption is often implied that no further training is needed to do an adequate job. I remember the case of a young woman referred to me for testing who complained bitterly about having lost custody of her two children. However, among many documented infractions, she had punished her six-month-old daughter by throwing the child across the room in a fit of anger.

Assuming that adult humans have been largely stripped of their instincts, they certainly *have not* been left defenseless. In the absence of instinctually guided behaviors, we are the beneficiaries of other, possibly more adaptive, attributes. Among many animals, but most especially humans, thinking and learning have become highly refined, perhaps taking the place of instincts, especially when it comes to complex social behaviors, like knowing when it is safe to walk up to a group of strangers and when it is not. From an adaptive standpoint, referring only to humans, as more and more neurological systems evolved to facilitate thinking and learning, those new systems could fulfill their optimal potential only if there was a concomitant release from instinctual imperatives. Instincts require individuals to behave in certain exact ways; but learning and thinking provide us with the opportunity to go well beyond patterned behaviors. In all likelihood, humans may be of all animals the *most* dependent on learning, so the assumption that we have large stores of innate knowledge may be folly.

If the importance of instincts declined among humans, what might have taken their place? The answer is "innately guided learning."[3] According to James L. Gould of Princeton University and Peter Marler or Rockefeller University, our learning is *heavily* biased and guided by innate, genetic imperatives. Thus, we are not obligated to behave in certain ways in specific situations, but our learning and thinking are nevertheless biased by genetic and biological factors that strongly influence the ways in which we can learn (and therefore think). Even the types of information we can process are biologically based. For example, our ability to categorize may depend on innate properties within our brains. We may have a built-in tendency to group objects together, to group people in the same category, and so on.

In the absence of strictly dictated instinctual behaviors, many animals are innately guided in how they learn and process information. In other words, specific behaviors are not mandated, but how we learn and think are heavily influenced by genetic heritage. As we have seen before, Fido lives in a much richer world than we do when it comes to smells. His learning is much more influenced by odors than ours. He can remember a person he met six months ago by the person's odor. Some birds can learn and later recall the hiding places of hundreds of seeds. We who are more intelligent than Fido or the birds, however, would find such learning tasks very difficult or impossible. What lessons can be easily acquired by some species are often impossible for others. Brain biology and genetics therefore establish limits on what we learn and even dictate what we will pay attention to and what we will ignore.

One recurring theme when it comes to the brain is that some types of information have more importance to us than others. Thus, although we experience the world as if we are objective observers, objectivity is an illusion. What we attend to is neither objectively determined, nor is it chosen randomly or impartially. Natural selection chose *what information we could process and what we could ignore.* Thus, our most "objective" efforts at thought are strongly biased by genetic and other factors.

TWO THINKING SYSTEMS

Consistent with what has already been stated, the brain is a complex information-processing system that involves many ways of accomplishing a variety of tasks. Like "memory" and "learning," therefore, "thinking" also encompasses a multitude of functions. At present, no one knows exactly how those functions are accomplished or all the places in the brain where those processes might arise. Thus, just as in the case of "memory," researchers have given generic labels to different "thinking systems." There is general agreement that humans manifest at least two *types* of thinking, and these categories have been given different labels by different scientists: "unconscious and conscious," "procedural and declarative," "automatic and controlled," "explicit and implicit," "reflexive and reflective," to name but a few.[4] In this book I will stick with the labels "reflexive and reflective" because the term "reflective," at least, describes the thinking process in question with a minimum of confusion. Before this section ends, however, I will further clarify some issues regarding the term "unconscious."

Reflexive Reasoning

Reflexive reasoning unfolds when we are correct in our thinking without any sense that we are even trying. A good example of reflexive reasoning is demonstrated in the activities of reading, speaking, or listening to others. Language comprehension relies on a vast storehouse of rules and the quick, nonconscious application of those rules.[5] Reflexive thinking, like that demonstrated in reading, occurs automatically once reading has been learned, and the complex rules that underlie reading cannot be fully or consciously explained, even by the best readers.

Consider the very simple sentence: *The despondent man walked slowly toward the cliff.* Although we can readily interpret each single word in this sentence, we simultaneously draw a number of inferences that are *not* stated. We effortlessly piece together the fact that the man is despondent and that he is walking toward a cliff. We also draw upon our knowledge about cliffs. We know that they are high places that drop off suddenly. We know that they can be dangerous or lethal to those who go over the edge. Thus, we quickly understand that the man might be suicidal, although that is not stated in the sentence. We can quickly analyze this sentence and draw a number of inferences, and we would be correct in all likelihood. We can do so almost instantly without any sense of struggle.

Now consider yet another example, illustrated by the following words: "Know go they can that the or lethal dangerous those to be who over we edge." Although this group of words makes no sense as it is written, it contains the same words as a sentence in the previous paragraph. This illustrates that when we write or speak, we effortlessly put words together so they make sense. Yet, do you recall the day in school when you learned about proper word order? Was there a lesson in which we learned how to make sense when speaking?

Even though many of us may never have had a class in physics, we intuitively understand why the board of a teeter-totter must be centered. Yet, how many of us could "prove" that the board has to be centered in order for the teeter-totter to work? This example illustrates the concept of "intuitive" physics.

All of these examples are types of "reflexive" reasoning, and as humans, we are overwhelmingly more adept at these processes. Yet, the rules that underlie this form of reasoning are far more complex than the rules that govern even complex math problems.[6] Most of us can read with little difficulty, but as the previous examples illustrate, reading comprehension and writing require a *vast* storehouse of rules in comparison to even complicated subtraction problems, yet many more of us would fail with the subtraction problem than fail to arrange words into an order that makes sense.

Reflective Reasoning

Many terms have been used to describe the ability to solve problems analytically, one of which is "reflective."[7] This is the type of thinking we normally associate with the word "thinking." It involves a conscious, deliberate attempt to solve problems. Although this mental capacity is often depicted as the sine qua non of human intellectual functioning, we do not come by this talent easily. It requires a purposeful effort and often many years of training. Furthermore, even individuals who have been highly trained in reflective processes may be unable to apply their knowledge in all areas.

Now, I will digress momentarily to consider a type of nonconscious reasoning, one that has been termed "the unconscious."

THE UNCONSCIOUS

We saw earlier that thinking has been described by many researchers as a dichotomy, and clearly, the terms "conscious" and "unconscious" are ubiquitous in our language. Thus, why did the experts feel the need to devise a new labeling systems when existing labels were already available? This is not a simple question to answer, but one that needs to be addressed. The term "unconscious" was popularized by Sigmund Freud who wrote extensively about "unconscious processes" (although others before him recognized the existence of "unconscious" phenomena).[8] The concept of "the unconscious," as Freud envisioned it, was a theoretical description of the process whereby forces of a *sexual, aggressive,* or *hostile* nature were kept at bay. Essentially, potentially destructive impulses were kept in check in "the unconscious" (although they nevertheless exerted their influence on behavior from time to time). Many scientists since Freud have taken issue with his theory, and thus, many contemporary researchers refrain from using the term "unconscious" because it carries highly specific connotations.

What really muddies the waters, though, is that Freud based his theory on observations of behavior, and he was an extraordinarily sage observer. Thus, he observed many phenomena that still need to be explained.

According to Freud's view, human conscious processes were "small and insignificant" compared with the vast store of mental processes that occurred outside of awareness (in "the unconscious," in Freudian parlance).[9] Since Freud, modern scientists have confirmed that most of the brain's functioning occurs out of awareness. One such scientist estimated that no more than 2 percent of the brain's mass is devoted to conscious processes.[10] The issue is not whether mental processes can occur out of our awareness, but rather, how or why. New brain research appears to be providing some answers to this critical issue.

Some mental processes occur out of awareness because they arise from brain structures that are older than the neocortex. For example, the cerebellum stores "motor" memories and some forms of conditioning, as we have learned, but it clearly has no means of producing consciously recallable information. Even most of the neocortex may be involved in processes that occur out of our awareness, in spite of the fact that the neocortex is the presumed source of our conscious reasoning abilities.

Some experts think that nonconscious processes speed up processing, though they have not been able to describe exactly how. In short, partly because of new brain research, some or many aspects of Freud's theory have fallen on disfavor.

Freud's theory hypothesizes that there was a great deal of energy necessary to keep unwanted impulses at bay. In light of what we know about the brain today, though, that seems unlikely. There is no doubt that the brain is the most energy-consumptive organ in our body, but that is probably because it consists of so many neurons. Brain cells take energy for metabolism, and there are far more brain cells in our heads than Freud could have envisioned. Furthermore, PET studies have shown that our conscious processes appear to be the most energy consumptive of the brain's processes, and this fact would be very difficult for Freudian theory to explain. Conscious processes, we know now, are probably energy consumptive because they require more areas of the brain, working in concert with one another. The brain's energy output is a fairly direct reflection of the number of cells that are active at any given time. It should be noted, though, that some modern researchers have used modern brain research to support Freudian theory, not discount it.[11]

The term "unconscious" has therefore not been used here because it carries specific, Freudian connotations. Nevertheless, we must still talk about those processes that occur out of awareness, so the term nonconscious, which is the preferred term for those who discount Freudian theory, has been used instead. "Nonconscious" does not carry specific theoretical connotations that are connected to a single theory. (I should add, experts do not take exception with the word "conscious," so we will continue to use it as a synonym for reflective processes.)

Regardless of the labels we choose, most experts concede that many of the brain's processes occur automatically, in the absence of our awareness. Perhaps surprisingly, even our "conscious" reasoning abilities are *heavily* influenced by what goes on behind the scenes. We will leave this issue of the nonconscious process and refocus our discussion on another issue of importance, one that pertains to our conscious, reflective thinking, which as noted is highly energy consumptive.

The Energy Guzzlers

Without a doubt, conscious mental processes require time and a large amount of energy, which raises an important issue. For millions of years, animals did quite well without reflective thought processes, so why did conscious, energy-consumptive processes evolve at all? Through my reading and study, I have encountered at least three plausible answers: First, we will never know. Second, as noted, our brains (and all others) are metabolically very active. They produce large amounts of heat. However, they can tolerate few changes in temperature, so the issue of cooling is crucial for brain survival. Thus, as an adaptation to keep the brain cool, one of our ancestors, *Australopithecus* (who came before *Homo habilis*) evolved a more efficient cooling system. Once a more efficient cooling system had evolved, the brain was free to grow much larger. Thus, a larger brain may have been an indirect result of a more efficient brain cooling system. This theory, first proposed in 1990, is termed the "radiator theory."[12] In sum, our intelligence may have been an offshoot of another adaptation.

Another theory for the evolution of consciousness does not necessarily contradict the "radiator" theory, although it is certainly less colorful. Many scientists have observed that a *major* outcome of conscious processes is our ability to alter the environment. Only we humans have this capability to any great extent. From the standpoint of adaptivity, conscious thinking abilities may have been naturally selected like any other capabilities because they provided us with survival advantages. (However, this hypothesis does not explain why the brain grew so large to begin with. That is a source of continuing debate for those who study brain evolution.)

We cannot resolve this issue here, although it is interesting to contemplate. Regardless of why "higher level" conscious processes arose, I must be clear about one point: Thinking provided us with the unsurpassed advantage to imagine things differently from how they are and simultaneously created the ability to modify aspects of the world to suit us. All other animals must adapt to their environments. Even beavers, who are considered nature's "engineers," must adapt to very restrictive environments and modify their surroundings, based on instinctual drives. Unlike humans, beavers cannot choose to *not* repair a broken dam, nor can they live in all parts of the world.

Often portrayed as the pinnacle of evolutionary success, there is no question that our brains are the most complex thinking systems in existence. The term "complex," though, should not be mistaken for "objective," even though the latter term is often used to describe a type of human thinking—objective thought. It is more accurate to assert that human thought is typically "good enough" in light of an individual's needs or goals at any given time.[13] "Good-enough," though, should not be mistaken for objective. On the contrary, our thinking manifests many biases that *increased* our chances of survival. We will start here by quickly reviewing some points that were touched upon earlier.

Meaning Precedes Thought

Although our ability to think symbolically has provided us with many advantages, the process of assigning labels has a biasing influence over thought. The act of thinking with labels goes beyond merely recognizing the individual properties of an object. Look around your room, as I am doing now, and let your eyes find an object. I will look at my dusty stapler while you look at the object of your choice. Whatever you are looking at probably consists of different colors and textures. It has shape, possibly an odor, and different parts of the object may even be separately identifiable. Notice, too, that your object is separate from the background (or else you could not see it). The properties of any object are registered by your brain, relatively effortlessly and nonconsciously, and each separate property of the object is integrated by an "association" center within the brain to form a whole. You and I attach a label to an object *after* it has been separated from the background and labeled. Normally we see an object as a whole or part of a scene. We rarely notice the individual properties of an object (although we can describe each property, like color, if we are asked to).

By labeling an object, we go well beyond its mere physical properties. When I say, "There's a car coming down the street," the word "car" has taken wavelengths of light, sounds, shape, and movement, and coalesced them into an object: a car. Some properties are accentuated while others are ignored. If we were simply seeing various sights, sounds, and shapes, the object would be no more recognizable as a car than anything else. By applying the label, though, we have consolidated separate properties into what is essentially a summary or label. A label or summary, though, is never exactly equal to the actual object. Beyond changing our perceptions by affixing labels to objects, the act of thinking requires us to take several steps that also influence our thought processes.

Thinking Requires a Premise

Beyond giving meaning to the world around us, thinking is based on underlying assumptions; yet we usually go about the business of thinking with little or no thought about the assumptions on which our thinking rests.[14] Our thinking is analogous to what happens when we walk into a skyscraper from the street. There are many levels of foundation underneath us that we do not see and often do not consider.

When we think, we are forced to adopt a starting point of view before thinking begins. If I asked, "What do you think about the Russians?" you could not "objectively" answer the question because you would have to begin your analysis by making several assumptions. If you agreed to respond, you would have to rely on previously learned definitions for each word in the sentence. Those definitions are called upon automatically, "reflexively." In calling up the meaning of the words, you would also be drawing upon past learning, yet you would not necessarily question where that learning came from or its accuracy. Would you, for example, wonder about the word "think" and ask yourself, "When did I first learn that word?" "I wonder if it has more than one meaning."

Second, you would automatically start with some belief or idea about the word "Russians." Quite possibly, each of us holds many beliefs and attitudes about Russians, some of which are stored nondeclaratively (possibly based on previous encounters or years of media exposure to stories about the Russians). One, or more, of those beliefs would be invoked by the question. To begin "objective" thinking, therefore, we start with an assumption that may remain completely hidden from view.

Most premises that form the basis of thought remain outside awareness (which is the crux of "The Automatic Principle" from chapter 4). When we meet a stranger, each of us proceeds in a characteristic way, often based on previous encounters with strangers, previous encounters with people who remind us of the current stranger, or perhaps some recent event.

One of the most pervasive influences on our thinking is *context*. Our brains are highly attuned to what is going on around us, and a good deal of our thoughts are influenced by what is available in the immediate environment.

Thinking and Context

The concept of "priming" (whereby our behavior is influenced by previous experience, but outside our awareness) is critical to understanding our thought processes.

Assume that you had just been reading a newspaper article before meeting someone new. You had read an editorial decrying the crime problems of today, blaming unruly youth for these problems. Then, as you meet the person, you notice a tattoo and an earring in one ear. Without "thinking" about it, you might make some judgment about the person based on the newspaper editorial.

Or consider another example. If people observe a forest fire after a lightning strike, they readily attribute the cause of fire to lightning. However, when people see fire in a building that displayed warning signs about oxygen in use, such as in hospitals, they will often attribute the fire to the presence of oxygen. In either situation, fire is equally dependent on oxygen, yet oxygen is likely to be cited as a causal factor in only the second instance.[15] This illustrates that we often make assumptions on the basis of information that is *contextually* determined, that is, information that is readily available. This makes perfect sense if we think of the brain as a dynamic organ within a dynamic environment. The brain must constantly reposition itself in response to changes in the environment. For our prehistoric ancestors, context would have been the key to survival (and it may remain the key to modern-day survival).

These examples illustrate that our thinking is biased by what we might see around us, often without our knowledge that we are being influenced by this information.

Another common influence on our thinking involves memory. If some event is more recent in our memory, that event may have a greater influence on our thoughts than other memories, even memories that are really more germane. Thus, if I read a newspaper account yesterday about an airplane crash, I might reason that flying is very unsafe, even though I may have read an article two weeks ago that more objectively described the airline industry's actual safety record.[16]

In a closely related matter, much of our thinking is based on attitudes, many of which are learned early in life, probably through social learning, and stored as nondeclarative memories. By attitudes I am referring to personal biases or beliefs, insofar as they influence behavior. If I had a lifetime of being teased at the hands of an older brother, I might react (nonconsciously) to older males in ways that are different from all other strangers I meet. Each of us bases many decisions on a lifetime's worth of conditioning.

Many attitudes stem from social learning, which can be incorporated as an image; it can be quite literal, and if it was established early in life, it can be quite childlike or even immature. If an attitude was formed early in life and stored nondeclaratively, it probably reflects the level of cognitive development that was present when the attitude formed. If a boy of five learns a set attitude toward strangers, that attitude may be drawn upon throughout his adult life and may reflect the sophistication of the individual at the age when the attitude formed. Here we have part of the basis of "negative stereotypes." We can incorporate negative attitudes and never update them because they are stored nondeclaratively.

There are good reasons to believe that many attitudes are formed early in life. First, the brain learns most voraciously during its earliest years, presumably for survival advantages, and many of those lessons were incorporated through the processes of social learning. The young of all species have to learn as much as possible, as early as possible, to ensure survival. Further, cultural, religious, social, and language lessons come easily to the young.

Since many functions of the brain are automated, including thinking, underlying premises are drawn upon habitually and automatically, typically outside of awareness. (A "premise" is simply the starting point of our thinking. In the Russian example earlier, once we began to think about Russians, whatever bias we began with is our starting premise.) Like any other mental processes that are often used, whenever a premise is drawn upon frequently, it will have a greater chance of being drawn upon again. This is probably because the neural networks that underlie the premise have developed into a patterned response. Thus, the nonconscious attitude "All strangers are dangerous" will be drawn upon automatically when encountering strangers. That attitude may never be updated to reflect more sophisticated reasoning ability such as, "Some strangers are dangerous, some are not."

Rarely do we know all there is to know about the underlying processes that influence our thinking, yet few of us question our thinking, and it is rare that we know when or where underlying premises came into existence. We can demonstrate this to ourselves. The next time we begin to think about something, we can write down our assumptions, but we have to go beyond the surface. If I said to myself, "I think I'll go upstairs to work on my book," that statement reflects an intent, but underlying the intent are many beliefs. What do I believe will happen if I do not begin to write? What will happen if I do? Why do I want to do this? By asking ourselves a series of questions, we get a sense of the "underlying premises" that *might* be housed nonconsciously, but we can never know for sure if we are absolutely right. However, some ideas will seem much more plausible than others.

In the realm of social psychology, humans manifest so many biases that, in fact, some of the more common types have been given names. "Framing," to illustrate, refers to the phenomenon whereby we will provide different answers on a questionnaire, depending on how the questionnaire is worded. If surgery patients are asked to determine whether to undergo a certain procedure, they will make a different decision, depending on how the informed consent might read. If they are told they have an 80 percent chance of survival, they are more likely to elect the surgery, than if they are told they have a 20 percent chance of dying.

What we have learned up to now is thoroughly consistent with the brain's penchant for automating as many processes as possible. The brain forms attitudes and engages in thought processes as automatically as possible, questioning as little as possible. Different premises are automatically invoked for different situations. This is theoretically one way for the brain to conserve fuel, and speed up response time, and effectively interact with an ever-changing environment.

Although our thinking is heavily influenced by our physical surroundings, we are particularly influenced when other people are present.

Thinking in Society

In social situations, where we cannot pull out a formula to extract the "right" answer, we often fail to agree on such nebulous issues as, "What is going on here?" "Who said what to whom?" "Who did what to whom?" "How did so and so react to Jane's comments?" Ask five party-goers to describe an event, and you will get five different descriptions of what happened. Ask those same people to explain the *meaning* of the event and then see what develops. The attribution of meaning to social or interpersonal events immediately plunges us into murky waters.

One well-documented fact about human thinking is that we often feel certain that others see a situation exactly as we do, even to the point of applying the word "truth" to interpersonal or social situations, in spite of evidence that no two people describe a complex social event in identical terms.[17] We are likely to overestimate the degree to which we are accurate in predicting the behavior of others. Generally, we have a strong tendency to feel confident in our predictions about the behavior of others. Further, when we are wrong, we are likely to underestimate the degree to which our predictions fell short of the mark. Furthermore, since we typically have confidence in our conclusions, confidence, per se, *decreases* the likelihood that we will double-check our conclusions. In other words, confidence in our own thinking affects the meaning we attribute to social events.

This is an example from a recent experience: I was visiting the post office during the holiday season. I had been waiting in line like several other patrons. When I got to the window, I was told that I had filled out the wrong form for the type of service I wanted, so the clerk told me to go back to the table, fill out the right card, but return to her window without going through the line again.

After a few moments, I returned to the clerk's window as I had been told. In the meantime, another customer had entered the line, and what do you suppose he saw? He saw this guy go directly to the front of the line, apparently cutting in front of everyone else. At that point he started making some very loud and objectionable remarks.

Do you suppose he ever realized that what he *thought* he saw was not the "truth"? (That is unlikely, unless someone told him the real story, which is doubtful because he did not appear to be very approachable at the moment.)

We often form judgments based on limited information, and where other people are concerned, our information is almost always limited. Why? Because we never know what is in another's mind, and we never know all the past history leading up to an incident, yet our brains quickly formulate judgments all the time as if our judgments were not subject to error. This is probably because the brain evolved to experience its conclusions as accurate, as stated before, especially if those conclusions had been based on something that was seen, heard, or in some other way experienced with one of the senses.

If our prehistoric ancestors had not believed their eyes, they may have waited too long to respond to danger. Also, taking the time to check one's conclusions takes more time, energy, and motivation. Thus, when more resources are required, averaged over large groups of people, tasks that require more energy and time decrease the likelihood that the task will be undertaken, especially when someone believes that a "correct" solution has already been found.

Stereotypes, Traits, Stories.   Researchers are discovering that certain patterns of bias are common enough to be considered typical, if not universal. We act and think differently when others are present. There are three categories of "bias" involving the presence of others that are particularly common (one of which has been discussed): stereotypes, traits, and stories.[18]

The brain's goal, at any given time, is to perform adequately for the demands at hand and then move on to the next challenge. Stereotypical thought processes serve that goal admirably. Thinking "shortcuts," whether stereotypes or traits, provide important advantages. They conserve fuel by recycling previous solutions and therefore reduce the time it takes for us to respond. In fact, it may be impossible for the brain not to stereotype insofar as thinking must invoke some preliminary starting assumption; and symbolic thinking, whether with words or numbers, requires us to assign a symbol before thinking begins, but when we assign a symbol to someone or some event, we are "stereotyping."

Typically, stereotypes are based on what we *perceive* to be the most relevant characteristic of another person. We even rely on stereotypes and traits to "predict" future events.[19] Even when provided with evidence that predictions are likely to be wrong, we are typically overconfident about our predictions, ultimately resulting in some pretty bad ones.[20] In fact, the fields of psychiatry and psychology have really taken it on the chin because of the imprecision involved in predicting the future behavior of individuals based on stereotypes (diagnostic labels). Having said that, some explanation is in order, because at the beginning of this chapter, I asserted that some people do a great job at making predictions.

Scientists, including social scientists, when engaged in experimentation, make hypotheses, and based on a great deal of information about a phenomenon they are studying, they make predictions about the expected outcome of an experiment. This is very different from making a prediction about how an individual is likely to behave in the future. When psychologists or psychiatrists are called upon to predict the likelihood that an individual will or will not be violent in the future, such predictions are notoriously inaccurate. Why? Because no one can know, in advance, what situations the patient will be confronted with in the future or the patient's state of mind when an incident or provocation occurs. (In spite of a terrible track record at predicting future behavior, psychiatrists and psychologists are constantly called into the court to testify about individuals with the implicit understanding that "expert" testimony will tell the court something about future behavior.)

Predicting group behavior, as we have discussed in this book, is far easier and often more reliable. The reason is that individual idiosyncracies or variations of one person are canceled out by those of another. Earlier we considered "intermittent reinforcement" in the context of gambling. Although I cannot say that any given individual will stay at the slot machine because of intermittent reinforcement, I can say that intermittent reinforcement will keep people *in general* coming back far more effectively than other schedules of reinforcement.

In addition to stereotypes and traits, we routinely construct stories to explain life's vagaries.[21] That is, human consciousness seems to compel us to weave the events of our lives together into a story that has continuity. If I were driving to work today and was hit by another motorist, I would "make up" a story that explained why it happened. "I am always running late, so if I had been on time, that car wouldn't have been there at that exact moment."

Not only do we tie our lives together with a story, but researchers are discovering that we do so by using a number of *strategies* called "counter-factuals." In other words, our storytelling strategies are not random.[22] For example, we often recompute the odds of a past event. "I was so close to winning! Why didn't I pick seven instead of eight?!" (In reality, the probability of winning on a losing number, after the fact, is zero.)

When people are asked to attribute a cause for a negative event, when equally plausible, yet competing reasons are available, they select the most "blameworthy" action as the main reason (even when each contributing factor may have been equally responsible).[23] If someone is speeding down the highway and gets hit by a drunk driver who is not speeding, the drunk driver will probably get a disproportionate amount of the blame, even though the excessive speed may have been just as blameworthy.

Researchers tell us that we even make up imaginary people against whom to compare ourselves. Who can forget about all the "starving people" in "such and such" country as the reason for not wasting food.

Jumping to Conclusions. As a professional psychotherapist, I am intrigued by the process of conducting therapy, in light of what we have just learned. In many ways, the process of therapy is the process of weaving meaning into a story that has personal relevance for both a client and therapist. When a therapist and client sit together, they share a common belief about what *should* happen in therapy. One consistent finding about therapy is the importance of the relationship, and part of the relationship dynamic is influenced by what the therapist and client believe is therapeutic. In fact, I would venture to guess that much of what goes on in therapy has never been empirically demonstrated to be necessary, but if it makes sense to both the client and the therapist, it works.

Furthermore, therapy is one of those unique situations in which the entire process is highly dependent on knowledge about an individual, yet almost nothing is really known firsthand about the various situations in which the client might be functioning outside the therapy room. In accordance with both ethical and legal standards, therapists are not supposed to have a personal relationship with their clients outside the therapy setting, so knowledge of clients and their lives is usually very limited. This point is especially interesting if we consider what was learned about the vagaries of memory.

During prehistory, and even more recently in recorded history, priests and other healers resided in the community. They lived and worked side-by-side with the very people they were entrusted to help. Now, in contrast, any contact between a therapist and client outside the therapeutic setting is considered an ethical violation, partly because "personal involvement" is thought to decrease "objectivity." Thus, ironically, therapists today may be forced to "make sense" of a therapeutic encounter by inventing information to explain events, especially if those events occurred in the distant past.

No Checks and Balances.   Although human beliefs and conclusions are often fervently embraced (and rarely questioned), thinking processes can result in errors or incorrect generalizations. *Yet in spite of the brain's potential for error, the brain has no built-in or automatic methods for checking its conclusions.* Since much of the basis of our thinking is "learned," we can learn that anything is accurate, if that is what we have been taught. As far as the brain is concerned, "two plus two equals four"—as a premise—is no more logical or illogical than "two plus two equals five." The only thing that separates a good idea from a bad one, from the brain's point of view, is feedback from the environment—such as a teacher's red marks all over a math test, or the collapse of a house built on faulty engineering principles. "Magical thinking" is no less credible to the brain than "scientific reality," unless the brain has been taught that one "reality" is better than another.

The ability to believe anything, in fact, is the basis for religion and cultural practices. This is not to say those beliefs are right or wrong, but rather, to assert that the mind can embrace many concepts with full conviction, even convictions that contradict one another.

The brain's ability to accept almost any premise is almost certainly a product of natural selection. If the brain had to analyze every situation anew and come up with a completely accurate assessment, the time and energy involved in doing so would be prohibitive, as we have discussed.

If the brain had no inherent means of checking its own conclusions, how could it have survived? The group. Most humans live in groups (today, our primary groups are families). If you live and work in a group or groups, as most of us do, it is the company of others that keeps our ideas in line, that gives us a "reality check." Since the human is a social animal, and since all social animals live in groups, there was really no reason for the brain to have evolved mechanisms to check the accuracy of its perceptions and thinking. To the contrary, it was important for people to believe what their brains told them, so the everpresent group made certain that thinking did not go too far astray.

Before moving on, let me emphasize that humans are not inexorably doomed to make horrendous thinking errors or silly generalizations. People can and do question their assumptions under certain circumstances, especially when they are instructed to do so. I used to think it was absurd for judges to tell jurors to disregard some piece of prejudicial evidence, but simple instructions about how to approach a problem often assist us to be less biased. On average, though, checking out our conclusions is not the typical response.

BRAIN STUDIES

Much of the research cited in this chapter comes from the field of social psychology and sociology. The term "thinking" is not a precise term that most scientists favor. People who conduct neurological research, for example, study specific brain functions, often relying on modern technology like PET scans. The researchers then study what part of the brain might consume more energy when a subject is thinking about a specific word. Brain research techniques that can study human beings in complex natural social situations, however, do not exist. Research technology has not progressed to the point that people can be studied in the real world, doing what real people do. Thus, much of what is known about "thinking" is inferred from *behavioral* studies, and much of that research is being conducted by social psychologists, sociologists, and others who study human behavior in natural settings.

Because thinking encompasses many different functions, there is not a direct correspondence between different labels, like "reflexive" thinking, and specific structures or regions within the brain. However, some key brain regions have been identified that are undoubtedly essential for some of the processes we have been discussing.

Virtually all modern inventions emerged after the time of the growth of the cerebral cortex, so it is considered to be the part of the brain that provided us with our high degree of inventiveness. The neocortex is a pretty large area, though, as we have seen, comprising perhaps 70 to 80 percent of the brain (although not all of the neocortex is involved in all thinking tasks).

The frontal lobes of the neocortex comprise about 20 percent of the brain's mass. It is the frontal lobes (and in immediately adjacent areas) that contribute heavily to symbolic language and our conscious ability to concentrate, reason, and solve problems. Working memory is also believed to be heavily dependent on the frontal lobes of the neocortex. Further, the ability to plan and execute logical, goal-directed behavior is also attributed to the frontal lobes, specifically, the prefrontal cortex.

The frontal lobes, in conjunction with the limbic system, also contribute to our humanness. These systems function together, along with the brainstem, to support *emotional behaviors* and the many functions served by emotions, including attachment to others. Because of the ways our emotions influence our thoughts, we have to conclude that both the prefrontal cortex and the limbic system have a strong influence on thinking that occurs in social or interpersonal situations.

Language, too, depends on the neocortex to a large extent (although some language functions may rely on other areas of the brain as well). The typical functions we associate with language, however, such as the ability to read and speak, the ability to recognize a written word, or the ability to recognize an object and attach a label to that object—each of these separate functions—are heavily reliant on areas within the neocortex.

Once again, even with our "most human" abilities, we see the networking principle illustrated. Highly complex mental activities, like reading and planning ahead, or responding to an emergency, cut across regional and possibly evolutionary boundaries, drawing upon nuclei from widely separated regions within the brain.

Much of the crux of the foregoing presentation is condensed and summarized in the following principle:

PRINCIPLE 12: THE THINKING PRINCIPLE

The brain's large "new cortex" is critical for those processes we call thinking and reasoning. In turn, thinking has provided us with the unparalleled advantage to change our environments in whatever ways suit us. Nevertheless, most of the brain's size is devoted to nonconscious, automatic processes that are heavily influenced by genetic, biological, and social factors (of which we have little awareness).

We can now apply what we have learned about thinking and consider ways of putting that information to use, to understand ourselves better and perhaps solve some problems.

IMPLICATIONS FOR UNDERSTANDING BEHAVIOR

Thinking about Thinking

As far as anyone knows, we are the only creatures that can "think" about our own thinking. We may ask, "Why bother?," but there are a number of advantages when we do so. First, thinking about thinking is one of the best ways to spot errors in logic and generate alternative solutions. Focusing on our own thinking is also a means of improving it. Just as we can exercise our bodies, we can exercise our minds with good results.

Also, for individuals who tend to be rash or impulsive, learning to think about their thinking often changes the locus of control within the brain. Any change in thought or action calls upon a slightly different configuration of brain cells, so by thinking about thinking, people become less emotional, more "cerebral." Moving the control of the process to the frontal lobes often results in a greater sense of self-control.

Based on my professional experience, one of the most important characteristics I look for in healthy functioning adults is some indication that the individual understands that perceptions and beliefs *can* be biased or mistaken. I have found that teaching clients to analyze their own assumptions is a worthwhile goal.

I do not wish to imply that any of us should lose confidence in our judgments. People who incessantly question their judgments, often deferring exclusively to the judgments of others, may also end up with serious problems. I am suggesting that individuals who *always* believe their conclusions and never test their assumptions might also have problems. There is no hard-and-fast rule, but there certainly is enough evidence to suggest that our thinking processes can be fallible.

Making Change

Our ability to think has provided us with the advantage of thinking about numerous *possibilities* and following through with change. We can imagine what it would be like to be wealthy and then systematically plan ways of making money.

Furthermore, we can contemplate changes in advance and consider alternatives and possible outcomes. We can even fantasize (imagine) about some changes we do not wish to make, to avoid mistakes or avoid the mistakes of others. We can consider what it might be like to spend a night in jail and determine not to engage in illegal activities because the risks are too great.

Thinking is like walking, swimming, or any physical ability. The more we practice, especially with the help of a coach or mentor, the better we get. We can actually practice being better thinkers and, therefore, fine-tune a capability that already provides us with such tremendous advantages, not only over other species, but over other people who are less-skilled thinkers.

Changing How We Feel

Thinking, emotion, and mood have tremendous reciprocal influences on one another. If I wake up in the morning feeling sad, I can use my thinking to counter my sadness. I can determine to get busy, plan out a series of things to do, carry through with my plans, and in short order change how I am feeling. (Not everyone can do this all the time, but the point is that thinking exerts major influences over feelings and mood, both directly and indirectly.) Through thinking and determination we can do a number of things that might change how we are feeling: from doing volunteer work, to reading a book or going for a brisk walk. Our thinking has enormous influence over mood and feelings, and we can harness the power of thought to change not only the world around us but our inner world as well.

Escape through Reading

We humans are the only species who have the use of written and symbolic language. Through reading, we can go anywhere in the world, we can go to any time, past or present. We can go to different planets or worlds that exist only in the imagination. We can learn how to do anything that has ever been done or look into the minds of the famous and infamous. There are practically no limits.

Setting Goals

We can look into the future, decide where we want to be five years from now, and greatly increase our chances of getting there. From our five-year goal, we can outline a series of steps on how we are going to get there, much like we can plan a trip with a road map and know with a high degree of certainty that we will be in a certain place at a certain time. There are possible pitfalls to this, but overall, the ability to plan ahead has provided us with an asset that is often taken for granted, but one that is nevertheless remarkable, considering all other species, including our first hominid ancestors, did not have this capability to the extent we do. (Entire books have been written on goal setting and planning, on various scheduling techniques, so we can use planning techniques to our benefit whenever we want.)

Social Perception

We have learned that our thinking is heavily biased within social settings. Just knowing that, however, can increase our perceptiveness. Just knowing about various brain biases can improve our thinking. If I know there is likely to be more about someone than meets the eye, that knowledge can give me the impetus to ask more about the person. We can do a lot to counter our brains' built-in biases just by knowing those biases exist and deciding that countering them is important.

Unraveling Our Nonconscious Thinking Processes

Nonconscious processes, by definition, are those that are not accessible to our conscious scrutiny. However, there are indirect ways we can study nonconscious processes. One of the best ways is for us to look for recurring patterns. We all repeat behavior patterns over and over that seem to unfold with a life of their own. However, we can study those patterns consciously, reflectively, and devise hypotheses about why they keep occurring.

If you have ever kept a diary for a long period of time, you may have noticed that certain themes kept emerging, perhaps in spite of a resolve to change some patterns. When patterns emerge in our behavior, we can glimpse into the world of nonconscious processes that are probably running on autopilot. With our awareness, we can pick new strategies. I should add, recognizing and changing nonconscious processes is simpler to write about than actually accomplish. A good therapist is often indispensable for helping people to see nonconscious patterns that are highly ingrained. A therapist can be extremely helpful for providing support and guidance in both identifying and helping to change some of these patterns because we often fall into old traps before we see them coming.

Storytelling

We are better "reflexive" thinkers than analytical problem-solvers, and we can more easily put words together into a sentence than solve a complex math problem. We also have what appears to be an innate tendency to weave our daily lives together into stories that have continuity and an overriding theme. Our propensity to make sense of our lives is probably a by-product of human consciousness.

Because we weave our life experiences together into stories with personal relevance, and because we evolved in small groups, one of the most powerful ways of influencing our lives is through involvement in small groups. Through group rituals and interactions, particularly those in which we actively participate, we can exert tremendous influence over our lives. Group rituals are the basis of religious practices, the basis of self-help groups, and the basis of some of our most cherished cultural practices, like marriage.

In the realm of professional psychotherapy, groups have long been known to be very powerful forms of intervention, although individual therapy still appears to be the most common form of changing behavior, probably because it is relatively difficult to form therapy groups. People are often intimidated by the idea of talking about their problems in a group, particularly in comparison to sharing those problems with a single therapist. Nevertheless, the power of the group cannot be overstated, and our thinking may be ideally suited for small groups, so anyone who wants to change behavior or reach goals might consider a small group for making changes.

We can also form our own groups. Self-help groups have had a very good track record, and one reason is probably because of our social heritage and the ways in which our brains respond to a social context.

CONCLUSION

Seventy to 80 percent of the brain is made up of the new cortex and the underlying cerebral hemispheres, which collectively have provided us with a unique set of abilities that are not believed to exist in any other species. In spite of our enormous intellect, however, we have not yet succeeded in labeling all the things our brains can do. Thus, we have to rely on labels that poorly describe the many capabilities we have.

"Thinking" and "reasoning" capacities comprise a large group of heterogenous skills that are revealed through our behavior. Our "cognitive" skills include the use of written and symbolic language, the ability to plan ahead (for many years in some cases), the ability to delay gratification for a bigger payoff in the future, and perhaps most important, the ability to imagine or visualize things as being different from how they are. We can also "try on" new behaviors and contemplate solutions through imagination before taking dangerous risks. We can use our thinking capacities to determine what we want, we can imagine different ways of reaching our goals and then pursue our dreams. We can literally go into the future (in our imagination), see how things *can be*, and then redesign the present to reach our future vision.

We have explored some of the more relevant issues pertaining to human thinking functions, but only a fraction of what has been written about and researched. The study of human thought is a quickly growing specialty area within the broader field of psychology, so many books would be required to do the emerging research justice. However, some of the more important attributes of our thinking processes are embodied in "The Thinking Principle," restated here.

PRINCIPLE 12: THE THINKING PRINCIPLE

The brain's large "new cortex" is critical for those processes we call thinking and reasoning. In turn, thinking has provided us with the unparalleled advantage to change our environments in whatever ways suit us. Nevertheless, most of the brain's size is devoted to nonconscious, automatic processes that are heavily influenced by genetic, biological, and social factors (of which we have little awareness).

* * *

The next chapter will summarize a little of what we have been discussing and will provide some additional comparisons between humans today and our prehistoric ancestors. Finally, we will revisit some theoretical notions about evolution and see what those theories might reveal about our future. Since we are the only species that can consider the future, we might as well give it a try.

CHAPTER 11

THE HUMAN MIND, THE GROUP MIND

We have now considered twelve general principles that were carefully written to reflect current scientific thinking about the brain and behavior. Although each principle provides insight into a specific area of behavior, collectively, they may provide a powerful model for thinking about ourselves and our society.

By all accounts, the brain reached its current size very recently in prehistory. The emergence of today's brain in a modern body may be no more than fifty thousand years old, although humans with genetic make-up comparable to our own may have first appeared two hundred thousand years ago. This is significant because we are essentially a *new* species.

In spite of our recent emergence, it is likely that the brain reached its current size before the invention of written language, the discovery of how to use metals, the creation of art, the domestication of crops, and the development of large urban centers. This is an important point because it reveals that our ancestors continued to survive through hunting and gathering for thousands of years *after* the brain reached its full size, so brain growth, alone, did not compel an immediate change in lifestyle (or perhaps it compelled a change, but not one that materialized quickly).

The brain's most recent growth occurred within the social confines of small hunting and gathering groups. Those groups probably numbered fewer than sixty individuals, although group size was variable, depending, in part, on the availability of food. Group size is an important issue because it provides clues about the social conditions under which the brain evolved, suggesting that we are social creatures who are dependent on close, interpersonal contacts within small, stable groups.

Just as we are social creatures, our brains are social brains. Human infants cannot survive without the help of others for the first several years of life, and during that time, infants cannot remain impervious to the influences of others. During the earliest years of development, we learn important language and cultural lessons through continuous contact with others, usually without any conscious intent to learn. This type of learning is heavily influenced by imitation and is not necessarily dependent on interpretation or thoughtful analysis. Even after humans reach maturity, they remain social animals for life, and one important aspect of social living is emotional expressiveness.

At birth, infants communicate through displays of emotion, and the number of their emotions increase within a few months after birth. Each emotion probably serves one or more vital purpose. Emotions provide us with a primary form of communication long before language develops. Emotions also serve the function of binding relationships and groups together; in addition, the ways we express our feelings, or withhold them, provide us with distinctive marks of individuality. Finally, emotions help to orient us to important events in the environment. They also reward us, punish us, and motivate us.

In addition to being social and emotional creatures, we have other characteristics that strongly influence our behavior. The very makeup of our brains influences how we process information. Our brains are extraordinarily complex networks, made up of billions of neurons and a much greater number of synaptic connections. As a network, our brains manifest two important properties that influence how we process information. First, our brains are a conglomeration of brain structures that originated during different periods of evolutionary history, with the consequence that we have different ways of learning, memorizing, and thinking.

Second, our brains rely on highly specialized clusters of neurons that are dispersed throughout the brain. For any given mental task or behavior, many cell groups are required. Therefore, behavior and mental processes are rarely (if ever) the product of a single group of cells or a single area within the brain. Beyond the fact that our brains are highly complex networks, they have yet other properties that influence how we process information.

Our brains, unlike computers, are biological. Their information processing is dependent on a constant supply of fuel and oxygen, yet our brains can store neither. Moreover, information processing is both electrical and biochemical in nature, relying on millions of brain cells, all of which results in processing speeds that are relatively slow. Therefore, problems of greater complexity or *novelty* require more time and more energy. Since the brain evolved in a dangerous world, the brain demon-strates a careful balance between processing *speed* and *adequate* problem-solving. Unlike a computer, the brain's method of processing information was influenced by the forces of natural selection. Thus, neither accuracy nor speed are among the brain's greatest strengths. Stated another way, our brains are rather slow when it comes to processing certain types of information, and even our best problem-solving efforts can result in faulty conclusions or failures.

As humans, we often interpret the world and store our experience through the use of symbols, either words or numbers. The attachment of symbols to real-life events, though, can *change* our perception of those events. For this and the reasons just discussed, information processing is always biased, yet we experience our perceptions and conclusions as being "accurate." (It was also hypothesized that the need to experience our perceptions as accurate was a naturally selected characteristic.)

Because our brains are biological entities, they *change* over time. They grow and learn, which means that all information processing does not unfold evenly throughout the life of our brains. The earliest years of brain development *restructure* our brains to some extent, readying them for the future. This "restructuring process" unfolds in response to genetic dictates and environmental demands. Once the restructuring process has been completed, during critical periods, each experience will have left an indelible mark on our brains (and behavior).

"Learning" and "memory" represent ways in which our brains record experience and store information. Our brains host a variety of learning and memory systems. Some types of learning are highly localized within our brains (limited to small areas); others types of learning are spread widely and stored in component parts. In short, our brains have a variety of ways of learning and memorizing, and various labels have been applied to these different processes.

Classical conditioning is an ancient form of learning insofar as we humans share this type of learning with animals that have much simpler nervous systems. This type of learning can be established and maintained outside of awareness and is resistant to reason or logic. However, classical conditioning can be altered through counterconditioning.

Another type of learning is called operant conditioning, which includes the notion that much (if not all) behavior is goal-directed. Therefore, we can understand many behaviors by understanding the goals of others. In turn, one of the most reliable indicators of others' goals can be uncovered by looking at the aftermath of behavior, by understanding what "reinforces" behavior. Like classical conditioning, operant conditioning is considered to be an ancient form of learning, insofar as we humans share this type of learning with many other animals.

Our newest learning system, conscious, reflective processes, is attributed to our enlarged "new cortex," and conscious thought processes are considered uniquely human. Our large neocortex is thought to be the newest part of our brain, and it is the region of the brain that has grown the most during the last period of brain evolution. The neocortex provides us with those capabilities that are considered uniquely human, like the use of symbolic and written language, the ability to plan ahead, the ability to control impulsive behavior, and the ability to make substantial changes to our physical and social environments. In turn, the neocortex allows us to adapt to a wider variation of environments than any other mammal.

Even our conscious thinking abilities, however, are heavily dependent on nonconscious processes. In fact, most of our brains' functions are carried out automatically and nonconsciously. Thus, what we are aware of at any given moment is only a glimpse of a much larger picture.

HUMANS TODAY, HUMANS YESTERDAY

Although the following comparisons are theoretical, the discussion is firmly grounded on what we have learned so far, and some important inferences can be reasonably drawn.

The Environment

In spite of its complexity and compelling nature, the brain alone will ultimately tell us little about human behavior, if brain functions are considered in isolation from the past or separate from the environment for which brain functions evolved.

Like all living creatures, we humans are inextricably tied to the world around us for life itself. When it comes to environments, though, we are influenced by two environments: One is the physical world in which we live, including such factors as the air we breathe, the water we drink, and the types of dangers we face; the other is the social environment.

The physical world of our prehistoric ancestors was believed to be quite different from the world we inhabit today in many respects.[1] While we should not minimize this point, our hunting and gathering ancestors, and their *Homo erectus* ancestors before them, traveled widely throughout the world, just as we do today. Even when the earth's climate changed, humans appeared to have adjusted quite well to those changes. Thus, the available evidence would indicate that both prehistoric and contemporary humans were, and are, extremely *adaptable*, in regard to finding ways of surviving in all parts of the world. People have survived in most parts of the world for thousands of years; and in some places, like regions of Africa, humans have successfully lived for close to two million years (*if* we can include the successes of our closest ancestors, *Homo erectus*).

Our success as a species is inextricably bound to *at least* two character-istics—our large brains and our ability to cooperate with other humans in small groups. Groups (society), therefore, as noted, constitute another type of environment on which we are dependent for survival.

Analyzing our ability to adjust to many social environments is more difficult than studying our adaption to the physical world. Not all experts agree about the extent to which we have successfully adapted to diverse social environments.[2] Some believe that we have created communities that are not healthy, communities that can foster mild mental retardation in children who, in all likelihood, would have had normal intelligence under more wholesome environmental circumstances.

Based on the information reviewed, it is safest to conclude that mod-ern humans have created social environments that are both healthy and unhealthy. It really depends on the specific environment to which we are referring. Modern humans (in contrast to our prehistoric ancestors) live under highly diverse social structures.

We know that we are social creatures and presumably our social nature was "naturally selected" because "sociability" provided our ances-tors, and us, with an adaptive edge. There is a corollary, however, to the natural selection principle, which we should consider more closely. Just as beneficial traits are chosen through natural selection, traits that are *not needed* are *not* selected. There are some possible exceptions to this conclu-sion, but overall, most traits and behaviors are presumed to be the result of natural selection. Therefore, if a group environment provided humans (or apes) with certain adaptive advantages, then *individuals* within the group did not need to evolve those characteristics. Thus, among humans, we had no need to be the fastest runners in the jungle, nor did we need claws to defend ourselves. The group protected us.

Just as we had no need to develop claws for protection, it is likely that our mental processes also showed a comparable pattern of evolution. In other words, there was no need for mental capacities to evolve that duplicated functions already being accomplished by the group as a whole. Thus, most of our mental capacities—thinking, learning, memory, and our emotional processes—can be understood best as functions that were naturally selected to operate within a small group culture. This conclusion is only tentative, but appears to be warranted by the evidence presented, although we may have to read between the lines to some extent. So, let's look at "the evidence."

First, the skull is quite crowded. There is little room for functions that do not (or did not) serve an essential survival purpose. In other words, space demands placed serious constraints on which functions were naturally selected and which were not. The brain was forced to evolve mechanisms that balanced competing needs. There was the need to conserve time and energy, on the one hand, and, on the other hand, there was the need for a system that could reach *adequate* solutions rather quickly. Because humans evolved in groups, the ability to form absolutely accurate decisions did not evolve because the ever-present group's consensus was good enough and the extra space needed to form absolutely accurate decisions would have required skulls that were much larger than the ones we now have.

Groups provided their members with certain advantages, like safety, nurturance, and the ability to indoctrinate children into the group's culture, but groups also influenced thinking and perception, insofar as human thinking is heavily influenced by the presence of others, as we have learned. Very importantly, group living also prescribed certain behaviors that protected individuals and the group by specifying the types of behaviors that were beneficial to the group as a whole. (All this *can* be true for modern groups as well). The foregoing statements are not intended to imply that groups were infallible in their "group decision making." During human prehistory, and to some extent during modern times, a group consensus that was too aberrant could have led a group to extinction.

Second, we humans have attributes that cannot develop outside a group. We can learn language only through exposure to other members of our species. Our emotional communications are finely tuned to convey messages to others, and even good health and longevity are influenced by nurturing emotional relationships with others.[3]

Third, for millions of years, leading up to the evolution of today's contemporary brain, our primate ancestors were social creatures. The brain reached its full size and potential within a group setting. We have to go back millions of years to find a creature in our lineage that was *not* social. If we subscribe to evolutionary theory, it follows that our brains are different as a result of social influences, in contrast to the brains that would have emerged in the absence of social forces.

Fourth, although only individual traits are naturally selected, individuals who live within groups have a survival advantage over those that live alone. Longer life, in turn, increased the chances that an individual's genes would be passed on to others. Thus, individuals well suited to group life have an edge over individuals who are poorly suited to communal living.

Fifth, as we learned, there is little evidence of instinctual behavior among adult humans. Our ability to live and function within society is primarily learned, and our brains evolved mechanisms to accommodate social learning.

Sixth, our brains are characterized by critical developmental periods, during which our brains are organized for later learning. Many, if not all, critical periods, and the subsequent restructuring processes, are dependent on exposure to other *people*.

Finally, our brains have no *automatic* mechanisms for checking the accuracy of their conclusions. To the contrary, our brains have a bias toward believing, even when conclusions are incorrect. Further, once we have reached a conclusion, we are far more likely to believe our conclusion than disbelieve it. Also, we are far more likely to seek out evidence that supports what we already believe than to seek out or spot evidence that contradicts our beliefs. This "bias-toward-belief" is likely to have been naturally selected, but if so, that selection process occurred within a small group milieu.

Consider what a "bias-toward-belief" might have been like for a single individual living in the wilderness. What if, while alone, a single prehistoric human believed that a particular campsite was safe, when, in fact, imperfect human eyes failed to see a dangerous predator in the trees? What if our ancestor's brain concluded that he could easily take on a large bear with only a club? These types of beliefs needed to be counterbalanced by an ever-present group. During human prehistory, given the lack of modern weaponry, we can assume that humans were rarely alone or far from other band members.

The brain can believe anything it is taught, and the capability of belief is necessary for assimilating the rules of culture and religion. If prehistoric parents said, "Stand up in the presence of your elders," children needed to do so without analysis. The ability to believe without question, though, needed (and needs) to occur within a social context, among other adults who are trusted, so that individual beliefs and conclusions can be counterbalanced by social forces.

The brain's ability to believe is not difficult to demonstrate. The editorial pages of our daily newspapers are filled with good examples, as are the pages of the "tabloids." Not long ago, I was standing in line at the grocery store, looking at the headlines of a recent tabloid. Next to me was a man who was also reading the paper. I asked him what he thought about a headline that had something to do with Martians. I was flabbergasted to learn that he believed the headline, with the justification that "they couldn't print it if it wasn't true."

My own opinions and observations notwithstanding, there is plenty of research to show that humans can hold widely diverging beliefs. Without looking too far, I ran across an article in a publication of the American Psychological Association. It stated that about 17 percent of the people who were polled claimed to have communicated with the dead. About 10 percent believed they had conversed with the "devil."[4] Now, let me be clear about one point. I am not assailing *a particular belief*. Rather, I am asserting that the brain is capable of believing almost anything, and we would not have survived in a wild world without the counterbalancing effects of an ever-present group.

Research on thinking has demonstrated that our brains do not invariably produce accurate conclusions when processing information. Rather, the results are most often "good enough."[5] Prehistorically, it is likely that individual thinking efforts were "good enough" because conclusions (and the behaviors based on those conclusions) were kept in check by other people.

Regardless of the actual importance of small groups (including the family) on brain development and functioning, there is consensus that we are social creatures.

We have already learned that groups consist of more than one person, and that three other dynamics are necessary: (1) interaction, (2) mutual dependency (working toward common goals), and (3) some type of structure. In addition to these dynamics, there are other characteristics of all groups that are helpful for us to consider.

First, groups can be characterized as "closed" or "open." Closed groups are ones that, as the term implies, are closed to new members. These groups can be made up of the same members from day to day, week to week, month to month, sometimes year after year. Families are good examples of closed groups. For long periods of time, the composition of a family can remain relatively intact. Military units are also "closed" as are many school classes, insofar as these groups remain generally stable over the course of a specified time period.

"Open" groups are ones in which group composition changes often, with members coming and going all the time. A church congregation is often "open," insofar as the exact makeup of a congregation may change from week to week.

Clearly, the distinction between "closed" and "open" is not hard and fast. No group remains the same forever. Even closed groups eventually lose members and recruit new ones or the group may cease to exist. The issue of "closed" and "open" is important because when groups have a more stable makeup, they develop different rules (culture) than groups whose membership is constantly shifting. For example, in open groups, where members can leave easily, conflict may be handled differently from closed groups, where members have no alternative but to devise a means of solving problems.

Groups are also influenced by the types of leaders and rules that operate within the group. Some groups are led by authorities with the power to make or change rules, or even censure individual group members, as in the case of families and the military. Other groups are egalitarian, where all members have relatively equal access to power.

Some groups have formal rules, such as bylaws; others operate informally, like a friendly neighborhood sewing circle where the "rules" are implicitly understood by each member.

Finally, when people interact with each other in a group, they generally develop a group "culture." As a professional psychotherapist, I have facilitated many therapy groups, and each group has certain properties in common. Even when people are forced to attend a group, as in the case of court-ordered offenders, it is remarkable that, in spite of members' objections to being in a group, over time, a group culture *always* emerges. Some members become leaders, others take on the role of "follower." Rules spontaneously emerge about such issues as punctuality, taking turns or not taking turns, about keeping silent or sharing openly, and so on. All groups develop a "personality" and a set of rules and expectations (whether those rules are spelled out or informally understood).

It is likely that both modern and prehistoric groups, while having many things in common with other groups, also had (and have) unique personalities. Therefore, the only thing that can be concluded with certainty is that groups develop a set of rules and expectations, and people within groups are influenced, to some extent, by the group culture. If members are not restrained by the group's culture, errant individuals are either forced to leave the group or the group's culture must change to accommodate the nonconforming member.

Culture

All societies and groups have beliefs, customs, and expected behaviors, that is, a culture. Acculturation pertains to the means by which beliefs and ideas are passed from one individual to the next over time. Among humans this occurs through learning, predominantly *social learning*.

Interestingly, cultures, like individuals, can be well adapted or poorly adapted to an environment. Some cultural practices can be so abnormal as to result in an entire group's extinction. In modern times, we have seen some extraordinary examples of nonadaptive cultural practices, like the mass poisoning of 913 individuals in Jonestown, Guyana, in 1978.[6]

Relative to human prehistory, we can only speculate about the reasons why different groups perished when they did, but clearly, cultures can be adaptive or nonadaptive, so group cultures were (and are) tested by nature. At present, anthropologists do not know why different groups became extinct. All prehistoric individuals eventually died, so how many died because of poorly adapted cultural practices and how many died from other causes will never be known.

Conclusions about Group Membership

Although millions of contemporary humans live in cities, large numbers of city dwellers, by definition, are "aggregates." Strangers on a bus do not meet the criteria for being a group. Similarly, the fact that an individual lives in a city does not mean that the individual is a member of any particular group.

One of the most striking changes from the time of our prehistoric relatives to the present is the degree to which society's basic social fabric appears to have changed. In contrast to our prehistoric ancestors, humans today live in highly diverse social environments. The range of diversity is nothing short of amazing.

Whereas all prehistoric humans lived as hunters and gatherers in small communities, today's humans belong to a multitude of different types of groups, and as modern humans, we are free to belong to large numbers of groups. (Although it could be argued that prehistoric humans belonged to more than one group, like smaller hunting parties within the larger community, there is no comparison with us.)

Just as importantly as our ability to join many groups, we can choose to participate in *no* groups. Individuals who do so, however, do not benefit from group membership nor are those individuals subject to the norms, rules, and pressures that arise from group membership. Even though individuals might interact with others on a casual basis, modern civilization has provided us with the freedom to escape group membership altogether, even in the most densely populated areas. (Naturally, young children are exceptions to this general rule during their years of dependency.)

What, then, might we conclude about ourselves, in comparison with our prehistoric ancestors, on the dimension of group membership? First, we moderns have the freedom to participate in a much larger number of groups, in comparison with our prehistoric ancestors.

Second, as adults, we moderns can choose what groups we belong to. Although we are all born into families, and may be born into religious and ethnic groups, we are free to leave our families during early adulthood, and free to abandon the teachings of our religious groups. In contrast, it is highly unlikely that our prehistoric ancestors could have chosen which groups they would belong to. Individuals were born into a band and likely remained there until death, or individuals moved to find a mate at which point they would have remained in a new band for life. Gender would also have determined which additional groups individual prehistoric adults might have belonged to. Men would have been hunters and women would have been gatherers and the caretakers of children.

Third, even though we may be members of more groups than our prehistoric relatives, it is likely that many of us spend *less* time in groups overall. It seems unlikely that our prehistoric ancestors would have spent any significant periods of time completely alone because of the presumed dangers of doing so, whereas we moderns can belong to many groups and spend much of our time alone. Modern humans can engage in work that isolates them from others, or work in environments that involve little or no interaction with other workers. We can work at home on a computer, work the night watch in a deserted building, or perhaps work as a night janitor and never see another soul during work hours.

Fourth, we moderns have access to a much broader variety of groups to which we can belong, and this issue relates to the concept of *consistency*. We can join many different groups, and it is possible to join organizations that have competing influences. Multiple-group membership opens up the possibility of diverse goals that may compete with one another. I can attend meetings with other people to help me lose weight (one group), and serve on a work committee at the office where another member always brings food that I cannot resist to our meetings. In contrast, prehistoric groups were essentially "closed" groups, in which membership remained stable over time, and the group, as a whole, had a single overriding purpose—survival. Thus, it is likely that prehistoric groups (one or two closed groups with consistent goals over time) probably exerted far different influences over group members and the brain than multiple-group membership in which individuals *may* be exposed to a wide variety of goals.

In conclusion, the brain evolved to accommodate a group structure. For hundreds of thousands of years, that structure was consistent in terms of relative size, purpose, and style of life. Today's groups are unlikely to provide modern humans with the same consistency of purpose, nor are modern groups likely to provide all humans with an equal amount of protection or access to resources as did prehistoric groups.

Child Rearing and Acculturation

Little is known for certain about prehistoric hunting and gathering societies, but some hypotheses are more likely than others. Because advanced weapons were not available and groups remained small, it has been hypothesized that prehistoric bands demonstrated a mutual dependency on one another, which also required them to subordinate individual needs to those of the entire community.[7] It is also believed that prehistoric bands maintained optimal population sizes for the protection of the entire community. This seems likely because, among contemporary hunting and gathering groups, there is a *predictable* relationship between food availability and group size. There are groups who have learned what size is optimal and have taken steps to remain at that size by means that we might find unbelievable.[8]

It is perhaps difficult for us to imagine the types of influences that small communal living must have had on our prehistoric relatives. Since beginning this project, I have often wondered if prehistoric groups knew how tenuous their existence was. Their entire community could go extinct if they did not preserve the group itself.

Within contemporary hunting and gathering communities, there have been recorded instances where infants were killed at birth to maintain optimal group sizes.[9] I am certain this fact will seem barbaric to many, but this example also illustrates the importance of preserving the group. Thus, with hunters and gatherers, we see a potential for paradox. On the one hand, they were forced by the demands of nature to keep their group sizes at a critical level, so even infanticide was not out of the question. On the other hand, they shared their limited resources with one another, they were mutually dependent on one another, and because group sizes were small, and bands were comprised of extended kinships groups, they were probably very close-knit.

Since humans lived as hunters and gatherers for most of human existence, whatever the shortcoming of that lifestyle (from a modern perspective), it was quite successful, *if* we measure success by longevity. To provide some perspective, if we look only at prehistoric humans who were genetically equivalent to us, the figures of comparison are one hundred thousand to two hundred thousand years for hunting and gathering, ten thousand to fourteen thousand years for all subsequent lifestyles.

In conclusion, it is reasonable to hypothesize that the brain evolved to accommodate hunting and gathering social structure. We can further hypothesize that communal life was ideally suited to the brain's needs because there was an ever-present supply of teachers to ensure that the "right" lessons emerged at the right time. Even if the biological parents died, other band members could have taken over critical parenting functions. Within any small community, we would expect that all children were exposed to all adults within the same community. Because of the longevity of hunting and gathering over the course of hundreds of thousands of years, it seems reasonable that a hunting and gathering social structure was well suited for the needs of all band members.

In contemporary society, there is not a uniform structure or style of child rearing, nor is there a "typical" family in which those practices occur.[10] Just as there are tremendous differences among groups to which we belong, there are significant differences in child-rearing practices. Moreover, families have been undergoing constant change throughout the history of this country, so there has never been a "typical" American family.

In the United States alone, we see children who live in two-parent families and single-parent and extended-family households. We see children who are raised in foster homes, institutional settings, mental hospitals, and even those who are homeless (some of whom may not have parents). This is only a superficial description of family life in modern America, and it does not even include the diversity found in other parts of the world. "Diversity" is the key word to just about all aspects of modern life compared with our prehistoric ancestors.

Although infanticide was probably practiced as a means of controlling group size during human prehistory for the good of the larger group, we moderns have comparable practices, but practices that generally are not based on the good of the community. Whereas civilization has decreased tremendously the threat we face from nonhuman predators, civilization has ironically increased the threats we face from members of our own species. "Indeed, modern history is characterized by increasingly efficient, systematic and institutionalized violence (e.g., the Inquisition, slavery, the Holocaust, the Trail of Tears)." Putting the recent past aside, it is estimated that five million U.S. children are victims of physical abuse each year, or minimally, they witness such abuse and violence.[11] Based on another source, of one million child abuse cases reported to various U.S. state or county agencies in 1994, 53 percent were neglected, "26 percent were physically abused; 14 percent were sexually abused, and 5 percent were abused emotionally; and 22 percent suffered other types of mistreatment." Of the one million cases reported, 1,111 children died.[12] As a country, the United States ranks last among eighteen industrialized nations relative to the number of children who live in poverty; that is, 20 percent of children live in poverty.[13]

None of this is intended to be an indictment, but only to illustrate how the conditions to which some brains are exposed today differ from the period of prehistoric hunting and gathering. Although no one would argue that hunters and gatherers were "wealthy" by today's standards, all hunters and gatherers shared resources relatively equally within their communities, and there is good reason to believe they were egalitarian in their attitudes toward one another.

The major conclusion in this section is that children today are reared under highly diverse social situations compared with a single type of lifestyle for prehistoric children. Thus, we would expect to see huge variations in child-rearing effectiveness, evidenced by enormous differences in adult competencies, and that appears to be the modern condition. What the effects of enormous disparities are cannot be determined at present, but the possible effects of those disparities are worth considering.

As we have seen, the primary learning mechanisms for assimilating cultural lessons is through reflexive and social processes. These are the types of learning that change our behavior, often in the absence of our awareness or intent to learn. "Watching" and "hearing" at the right time can exert powerful influences over children and adults. "Seeing is doing." Mere exposure to other members of our species is often enough for some lessons to take root for life.

Perhaps one of the most critical changes from the time of prehistory to that of today is the degree to which *outside influences* can impact the developing minds of today's children. During human prehistory, hunting and gathering lifestyles required large expanses of land. Thus, individual bands were often widely separated from one another. This suggests that children could have been exposed only to the influences of members of their own bands, with rare exceptions (if there were any exceptions). In short, it is all but inconceivable that prehistoric children could have been *indoctrinated* by *outside* forces. Further, careful indoctrination to the group's beliefs early in life and the lack of outside influences would have protected band beliefs from any serious external threats.

Although there is no evidence that prehistoric bands engaged in *widespread* warfare with others bands, occasional conflicts with other band would not influence the day-in, day-out acculturation of individual children within the prehistoric band. Again, then, the conclusion is that our brains did not evolve to contend with highly diverse influences through the most critical developmental years. For hundreds of thousands of years, no other social structure existed.

In contrast to our prehistoric ancestors, children of today are far more subject to influences from outside the primary socialization group (the family), and in the past few years, the primary source of out-of-family influence has come from television. By eighteen years of age, the average child in the United States will have spent fifteen thousand hours watching television, which is four thousand hours more than they will have spent in school—more time than was spent interacting with teachers, possibly more time than was spent talking with parents.[14] When I first read this statistic, I did some quick figuring, and fifteen thousand hours is equivalent to about 7.5 *years* of going to a full-time, forty-hour-per-week job, minus two weeks of yearly vacation.

In and of itself, the number of hours of television viewing does not begin to tell the whole story. Recent research conducted at the University of California, Santa Barbara, in conjunction with other universities, revealed that 57 percent of the television programs reviewed depicted scenes of violence, and one-third showed nine or more violent interactions. More telling, though, is that over half (58 percent) of the violence failed to show any painful consequences, and 47 percent of the violence failed to show that a victim was even harmed by violence.[15] When we consider the brain's ability to imitate and practice what it sees, accompanied by the inability to ascertain the "rightness" of those actions during the first few years of life, these facts take on a much more sinister significance. Through thousands of hours of television programming, children learn that violence is perfectly acceptable, and this learning becomes embedded nonconsciously as an attitude about violence and as a model for how problems are to be solved.

Finally, we should not assume that as children grow older they automatically become more adept at seeing through deceptive television messages. Two hundred teenagers in Missouri were surveyed to study their beliefs about commercials that were being aired on Channel 1, which is a broadcast system that provides televised information to schools.[16] Even children who had reached their teenaged years often mistook advertising as a public service announcement. Teens often failed to recognize that messages were being endorsed by paid actors, and some teens even believed that professional athletes were paying to be in a commercial (as opposed to being paid for the commercial).

In addition to greater amounts of time being exposed to out-of-family influences, modern children are with their primary socialization group (the family) far less than would have been the case for prehistoric children, who would have been with their caretakers (or other adults) twenty-four hours per day. In short, today's brain could not have evolved in response to modern conditions, partially because "modern" conditions are different today than just thirty years ago. One of the characteristics of modern life is the degree to which rapid change has been made possible by new technologies. As we have learned, human thought processes did not evolve to be highly analytical (as the brain's first response), but nonconscious processes are especially undiscerning. The brain never evolved to filter messages of questionable validity. Although humans can employ critical thinking skills, those skills must be taught, often through many years of formal education.

As prehistoric band society dissolved, the brain went from being exposed to a single group with a single overriding purpose to multiple group membership, with potentially conflicting goals and needs. Clearly, with some minor exceptions, the brain is now exposed to more external influences from outside the primary living group. However, since the brain evolved to accommodate the ever-present group, presumably within a single group structure, when that structure disappeared, an important pillar of stability *may* have been lost forever.

THREE DYNAMICS OF BEHAVIOR

Understanding behavior is like studying a school of fish. The school keeps moving and it is often obscured by the water. To understand behavior, we have to consider four dynamic forces: the brain, our physical environment, our social environment, and behavior.

The brain controls and influences behavior. Behavior, in turn, can alter the physical or social environment in which we live. By altering an environment, we can change the brain and behavior.

We can directly alter our brains intentionally or accidentally. We can modify our brains through medication or drugs. We can change them through surgery, or they can be altered by injury or disease. However, as we have seen, there are less invasive means of permanently changing our brains. We can change them through learning or thinking.

We make minor alterations to our brains every time we learn a new skill, learn a new fact, or try a new behavior. Among children, we can make substantial alterations, often quickly, by providing them with certain experiences during key periods of development.

We can make conscious decisions to change our behaviors. Even in spite of enormous fear, we can approach a spider known to be harmless, and our fears will eventually dissipate. Afterward, having altered our fear, we will have altered our brains as well (in a small way). We can consciously determine to try something new today. If we try new things often enough, we will establish new patterns of behavior, which establish new connections within the brain. However, when we change our behavior, we often alter our social or physical environments, or both. By altering the environment, we influence behavior, which in turn, influences the brain. This discussion illustrates a powerful concept—it describes dynamic forces that we can influence (to greater or lesser extent), and through those

244 BRAIN WAVES THROUGH TIME

dynamics, we can change our lives, achieve goals, and potentially change the world.

We have spent a good deal of time considering our past and present, but since we are the only creatures that can contemplate different future possibilities, we can now look at some potential outcomes relative to human evolution.

THE FUTURE AND HUMAN EVOLUTION

Theorists disagree about the rate of evolutionary change, but there is one evolutionary dynamic that has not been seriously challenged by the experts. A diverse gene pool ensures variability on which the forces of natural selection can operate. Given large human populations and the likely variability within the gene pool, it seems probable that we humans have many potential traits that are quite diverse. In theory, then, we can presume that there are extra genes out there somewhere awaiting the forces of natural selection to pick them. In the distant past, environmental changes are thought to have played a key role in natural selection by creating unique niches that needed to be filled through natural selection.

We moderns win the prize for making substantial changes to our environments (both physical and social), and there is no sign of slowing in sight. Thus, we *might* be contributing to our own evolution by creating new environments or contributing to the evolution of a new species that could supplant us. In either case, what characteristics might a "future, humanlike species" have?

Unless there are reductions to the world's population, humans of the future will have to adapt to enormous urban populations—*and we could be that animal*. We are extraordinarily inventive. Our history, to date, has demonstrated nothing about us, if not our creativity. Thus, one possible evolutionary scenario is that we will remain the planet's dominant species, with or without further evolution.

However, because of our penchant for making changes, it seems logical that there must be limits beyond which even we cannot go. As time goes on, we continue to invent more powerful means of changing the earth, and some scientists have asserted that we now have the mechanisms for making the planet uninhabitable, even for us. Through nuclear and biological technology, we humans have invented two types of weapons that could destroy all human life.

Barring a major world disaster of our own making, such as a nuclear holocaust or the release of biological weapons, all of us, collectively, are contributing to changes in the world climate and environment that could devastate the lives of millions. According to Paul Simon, a former U.S. senator, in the next fifty to ninety years, the world's current population of roughly 5.9 billion people is expected to double, at a time when the world's renewable water supply remains constant.[17] He also pointed out that as we gain in affluence, our per capita water use increases, and generally, our use of water has been rising much more quickly than the world's population. Thus, by 2050, many countries may run out of usable water.

In addition to the enormous potential for man-made disasters on a scale that was impossible prior to modern industrialization, modern society may be producing more individuals who are *less* socialized, which increases the chances that one (or more) of them will use the destructive technology that has already been invented. Perhaps as many as five million children are neglected or abused each year in the United States alone, and each of these children is a potential time-bomb.[18] In light of how the brain repeats past learning, this number is all the more ominous because it would suggest that succeeding generations will produce more and more children who are reared in unhealthy environments. Absolute numbers, however, do not tell the whole story.

Within the past few years, new terms have crept into our vocabulary, expressions so new that we may not find them in our reference books: "super predator," "serial killer," "sexual predator," "serial rapist," and "mass murderer." These terms refer to individuals who use their intelligence to prey on others. As a class of individuals, these people often appear to be devoid of empathy.

In spite of the warnings of experts, we will probably see more of these individuals, especially if current trends continue. When children spend an ever-increasing amount of time in the company of televisions, computers, and virtual reality environments, they are not learning how to develop emotionally based relationships with others. Although computer chat rooms provide the mechanisms for people to communicate with one another, there is no reason to believe, as yet, that chat rooms can teach empathic skills for others. In fact, communications through the Internet may reduce intimacy because it provides a mechanism for people to "interact" with others in ways that are impersonal, ways that demand less understanding among individuals. Thus, for some at least, computer technology may contribute to individual isolation, not decrease it.[19]

Empathy, as we have seen, is most often based on emotion and a personal involvement with others. Historically, our emotion-based attachments have held groups and communities together. That foundation is eroding for many groups. Without doubt, we can teach children that certain behaviors are wrong, but whether a knowledge of morality is enough to hold groups together without empathy for others remains to be seen. This is not to say that children who watch television have no empathy, but rather, television cannot teach empathy, in all likelihood, particularly if children are spending more time with electronic devices and less time with parents and other children. In and of itself, knowledge is unlikely to maintain groups and families, partly because children are now being exposed to so many conflicting ideas about what is right and wrong.

Our brains are not well suited to consider enormous numbers, whether those numbers represent people or brain cells. Our brains have a built-in bias for the simple and concrete, so the immediate needs of our families (which is more concrete) will take precedence over the needs of "the community," which is an abstraction, especially if that community is made up of numbers of individuals that are too large for the brain to comprehend. Thus, in the near future, we are unlikely to subordinate our individual needs, or those of our family, to the needs of the larger community, but for species survival, the needs of the larger world community may have to come first. Putting the needs of the larger community above the needs of the individual may seem absurd to us, but for hundreds of thousands of years, prehistoric societies did just that. The preservation of the group took precedence over the needs of a single individual.

In the distant past, no number of neglected children would have had any impact on the larger world community. The actions of a single individual or group would have had only a minimal impact on other groups because of small world populations and primitive technology. The story is different today. As we have already witnessed in the twentieth century, modern technology has endowed single individuals or small groups of people with the ability to affect the lives of millions.

In sum, two possible evolutionary scenarios is that we could orchestrate our own demise through the use of modern weapons, or through the combined effects of billions of people worldwide who, collectively, are modifying the planet in ways that are difficult to anticipate or measure. Clearly, there is no evidence that our brains have evolved since the last time so-called weapons of mass destruction were deployed. About sixty years ago, Hitler deployed poisonous chemicals in his attempts to exterminate millions of people during the Holocaust and the United States deployed two atomic bombs. It is unlikely to assume that our brains have evolved in such a short amount of time.

There is another evolutionary scenario that is far more positive than the extinction of our species or the death of millions of people. We could develop into humans that have the capacity to care genuinely for large groups of people. Some of us already demonstrate this capability, so the ability is not dependent on millions of years of additional evolution. Social learning provides the mechanisms whereby we could transform human behavior very rapidly. We could more fully develop our large intellectual capacities and become even more reasonable, developing into beings among whom logic and highly reflective processes are the norm. However, for this scenario to become a reality, *all* children would have to be subjected to positive early childhood environments, not to mention many years of formal education.

Logical, reflective processes and the ability to think are our newest skills, and among us, there have been extraordinary examples of individuals who have accomplished great feats. There are those who have eliminated diseases through the efforts of their thinking. There are those who have discovered better ways of feeding large populations, and there are those who have written great books that have positively touched the lives of millions. In short, we could realistically develop into a species of intellectually gifted individuals who use their intellect to focus on the improvement of the species and the planet.

We humans are subjecting ourselves to an ongoing experiment: We keep changing our environments, and then observing what happens after the fact. As more and more people are added to the world's population, the pace of the experiment accelerates. More people will inevitably transform the world to something other than what we see today, and human history supports the conclusion that we constantly change things to suit ourselves without knowledge of the long-range consequences. However, evolution itself is an experiment, so we as humans may be no better or worse off than all other species who have been subjected to the forces of natural selection.

Brain Research and Philosophy

As we have learned, for most of recorded history, beliefs about behavior and human nature have been studied within the domains of theology and philosophy. Historically, even scientific attempts to study behavior have been heavily influenced by religious and philosophical views. Even today, one issue of perennial importance is that of "free will." Does such a thing exist?

In keeping strictly with the principles presented in this book, we can conclude that humans have free will to the extent that underlying brain structures and previous experience are present to support a range of choices. However, if I am tone deaf, it is inaccurate to say that I have the "free will" to hum a middle "C" whenever the mood strikes me. Moving beyond this analogy, there is no reason to believe that other brain-based behaviors follow other principles.

Whether we are completely normal (whatever that might mean), or mentally ill (however that is defined), none of us is immune to what our brains tell us about "reality." Further, as we accumulate experience, which becomes stored as learning and memory, we often become locked into our ways of seeing the world. As a result, some people end up with beliefs that they can accomplish nothing. Others end up believing that they can fail at nothing.

All of us behave in accord with the beliefs we hold to be "true," and our brains do not routinely and accurately assess the validity of those assumptions. Thus, at the level of individual neurons, it appears as if *all* brains follow the same operating principles, even brains that are diseased. Once neurons have passed on their messages to us, we are highly influenced by the resulting picture, often with no knowledge of how effectively the information was pieced together.

* * *

The "modern" brain, by and large, is still the "prehistoric brain" in all important respects. In the area of thinking, the brain is still most adept at quick, global, reflexive processes. Our brains more readily process simple messages; our brains are more likely to believe than disbelieve what they see. Averaged over large groups of people, our brains still eschew sustained periods of mental concentration. Today, as in prehistoric times, behavior is largely governed by nonconscious rules and by beliefs about which we have little awareness. Habitual methods of perceiving and responding are the norm—whenever those responses appear even remotely to meet the demands at hand. Great contemplation or analysis is *not* normative, even though today's social world is far more complex than the societies of yesterday, and life in today's world requires us to consider the needs of people we have never even met.

Although our brains have numerous quirks, they are nevertheless marvels of adaptiveness and ingenuity. Our ingenuity, in turn, is often employed to change the world in which we live. As we change the world, though, we challenge further the adaptiveness of our brains. Then, as we adapt, we change the world a bit more. How long, we might wonder, can this progression occur?

Before we run out of adaptive strategies, we should use our incomparable reasoning power to learn more about the effects we have on our planet and one another. If we can send men to the moon, we can certainly use our giant brains as tools to both understand one another and preserve the planet on which we all depend for life itself.

NOTES

CHAPTER 1: THE PARADOX

1. Steven Zalcman, editor, "Neural Basis of Psychopathology," in *The Neuroscience of Mental Health, II: A Report on Neuroscience Research*, ed. Stephen H. Koslow (Rockville: U.S. Dept. of Health and Human Services, 1995), pp. 159–179.

2. Jo Thomas, "McVeigh's Pre-bomb Letters Reveal Troubled Man." *Rocky Mountain News* [Denver] July 1, 1998: 2A+.

3. Robert W. Lundin, *Theories and Systems of Psychology*, 3rd ed. (Lexington, Mass.: D.C. Heath & Co., 1985), p. 22.

4. Erin D. Bigler, Ronald A. Yeo, and Eric Turkheimer, "Neuropsychological Function and Brain Imaging: Introduction and Overview," in *Neuropsychological Function and Brain Imaging*, eds. Erin D. Bigler, Ronald A. Yeo, and Eric Turkheimer (New York: Plenum Press, 1989), p. 1.

5. Lundin, p. 10.

6. Bryan Kolb and Ian Q. Whishaw, *Fundamentals of Human Neuropsychology*, 4th ed. (New York: W. H. Freeman, 1996), pp. 4–5.

7. Stephen S. Carey, *A Beginner's Guide to Scientific Method* (Belmont, Calif.: Wadsworth Publishing Company, 1994).

8. Richard Cole, "Therapists Implanted Abuse Memories: Jury Says Sex Abuse Imagined," *Albuquerque Journal*, May 15, 1994: A1+.

9. James R. Averill, "Differences Between Anger and Annoyance," in *Anger and Aggression: An Essay on Emotion* (New York: Springer-Verlag, 1982), pp. 229–252.

10. Ibid.

11. Stephen B. Klein, *Learning Principles and Applications*, 3rd ed. (New York: McGraw-Hill, 1996), p. 207.   ↳ punishment

12. U.S. Department of Justice, *Special Report: Drunk Driving* (February 1988), cited in Stephen B. Klein, *Learning: Principles and Applications*, 3rd ed. (New York: McGraw-Hill, Inc., 1996), p. 207.

13. C. H. Patterson, *Theories of Counseling and Psychotherapy*, 4th ed. (New York: Harper & Row, 1986), p. 573.

CHAPTER 2: HUMANS IN PERSPECTIVE

1. Verne Grant, *The Evolutionary Process: A Critical Study of Evolutionary The-ory*, 2nd ed. (New York: Columbia University Press, 1991), p. 16.

2. Charles Darwin, *On the Origin of Species by Means of Natural Selection, or the Preservation of Favoured Races in the Struggle for Life*, cited by John G. Fleagle, *Primate Adaptation and Evolution* (New York: Academic Press, 1988), p. 9.

3. E. Peter Volpe, *Understanding Evolution* (Dubuque, Iowa: William C. Brown, 1967), p. 13.

4. Ibid., p. 14.

5. Ibid.

6. John G. Fleagle, *Primate Adaptation & Evolution* (New York: Academic Press, 1988), p. 1.

7. Stephen Jay Gould, "Exaptation: A Crucial Tool for an Evolutionary Psychology," *Journal of Social Issues* 47, No. 3 (1991), pp. 43–65.

8. Ibid.

9. Stephen Jay Gould, *The Panda's Thumb* (New York: Norton, 1980), cited in Robert J. Wenke, *Patterns in Prehistory: Humankind's First Three Million Years*, 2nd ed. (New York: Oxford University Press, 1984), p. 106.

10. Volpe, p. 16.

11. Grant, p. 16.

12. Volpe, p. 24.

13. Ibid., p. 25.

14. Tim Megarry, *Society in Prehistory: The Origins of Human Culture* (New York: New York University Press, 1995), p. 63.

15. Henry W. Nissen, "Axes of Behavioral Comparison," in *Behavior and Evolution*, eds. Anne Roe and George Gaylord Simpson (New Haven, Conn.: Yale University Press, 1958), pp. 183–205, cited in Carol R. Ember and Melvin Ember, *Anthropology*, 3rd ed. (Upper Saddle River, N.J.: Prentice-Hall, 1998), p. 24.

16. Robert G. Wesson, *Beyond Natural Selection* (Cambridge, Mass.: MIT Press, 1991), p. 284.

17. Volpe, p. 129.

18. Robert J. Wenke, *Patterns in Prehistory: Humankind's First Three Million Years*, 2nd ed. (New York: Oxford University Press, 1984), p. 54.

19. Ibid.

20. Ibid., p. 57.

21. Roger Fouts and Stephen Tukel Mills, *Next of Kin* (New York: William Morrow and Company, 1997), p. 55.

22. Michael D. Lemonick, "How Man Began," *Time* 143 (March 14, 1994), p. 83.

23. Fleagle, p. 8.

24. Ibid., p. 59.

25. Carol R. Ember and Melvin Ember, *Anthropology*, 3rd ed. (Upper Saddle River, N.J.: Prentice-Hall, 1998), p. 33.

26. Workshop by William H. Polonsky, "Social Connections, Health & Longevity" (Palo Alto, Calif.: Health Science Seminars, March 6, 1998).

27. Fleagle, p. 435.

28. Lemonick, p. 83.

29. Wenke, p. 54.

30. Richard E. Leakey, *Human Origins* (New York: E. P. Dutton, 1982), p. 12; Ember and Ember, p. 32.

31. Lemonick, p. 83.

32. Brian M. Fagan, *People of the Earth*, 9th ed. (New York: Addison Wesley Longman, 1998), p. 59.

33. Lemonick, p. 83.

34. Ibid., p. 84.

35. Ibid.

36. Ember and Ember, p. 64.

37. Lemonick, p. 83.

38. Fagan, p. 81.

39. Wenke, p. 100.

40. Ember and Ember, p. 70.

41. Lemonick, p. 84

42. Ember and Ember, p. 70.

43. Ann Gibbons, "Y Chromosome Shows That Adam Was an African," *Science* 278 (October 31, 1997), p. 804.

44. Ember and Ember, p. 79.

45. Michael D. Lemonick, "Stone-age Bombshell: Stunning Cave Paintings Turn Out to Be the Oldest as Well, Upsetting Some Assumptions of Art History," *Time* 145 (June 19, 1995), p. 49.

46. David Sloan Wilson, "Human Groups as Units of Selection," *Science* 276 (June 20, 1997), p. 1816.

47. Fagan, p. 17.

48. Ibid., p. 136.

49. Ibid.

50. Ibid., p. 155.

51. Ibid.

52. Ibid.

53. Wenke, p. 156

54. Ibid.

55. Lemonick, p. 82.

56. Bruce D. Perry, "Incubated in Terror: Neurodevelopmental Factors in the "Cycle of Violence," in *Children in a Violent Society*, ed. John D. Osofsky (New York: Guilford Press, 1997), p. 125.

CHAPTER 3: THE FIRST NETWORK

1. Robert J. Wenke, *Patterns in Prehistory: Humankind's First Three Million Years*, 2nd ed. (New York: Oxford University Press, 1984), p. 54.

2. Bryan Kolb and Ian Q. Whishaw, *Fundamentals of Human Neuropsychology*, 4th ed. (New York: W. H. Freeman, 1996), p. 39.

3. Lokendra Shastri and Venkat Ajjanagadde, "From Simple Associations to Systematic Reasoning: A Connectionist Representation of Rules, Variables and Dynamic Bindings Using Temporal Synchrony," *Behavioral and Brain Sciences* 16 (1993), p. 418.

4. James Bevan, *The Simon and Schuster Handbook of Anatomy and Physiology* (New York: Simon and Schuster, 1978), p. 60.

5. Neil R. Carlson, *Physiology of Behavior*, 3rd ed. (Boston: Allyn and Bacon, 1986), p. 93.

6. Ibid., p. 97.

7. Kolb and Whishaw, p. 41.

8. Nico H. Frijda, *The Emotions* (Cambridge, Mass.: Cambridge University Press, 1986), p. 165; Eric Halgren, "Emotional Neurophysiology of the Amygdala Within the Context of Human Cognition," in *The Amygdala: Neurobiological Aspects of Emotion, Memory, and Mental Dysfunction*, ed. John P. Aggleton (New York: John Wiley, 1992), pp. 191–228.

9. C. Thomas Gualtieri, "The Contribution of the Frontal Lobes to a Theory of Psychopathology," in *Neuropsychiatry of Personality Disorders*, ed. John J. Ratey (Cambridge, Mass.: Blackwell Science, 1995), p. 153.

10. J. Graham Beaumont, *Introduction to Neuropsychology* (New York: Guilford Press, 1983), p. 39.

11. Carlson, p. 86.

12. Bevan, p. 56.

13. Kolb and Whishaw, p. 45.

14. Ibid.

15. James D. Fix, *High-Yield Neuroanatomy* (Baltimore: Williams & Wilkins, 1995), p. 27.

16. National Institute of Neurological and Communicative Disorders and Stroke, *Know Your Brain* (U.S. Government Publication, Doc. HE 20.3502.B 73/6, 1987), p. 2.

17. Bevan, p. 56.

18. Carlson, p. 109.

19. Bevan, p. 62.

20. Ibid., p. 60.

21. Kolb and Whishaw, p. 48.

22. Ibid.

23. *Know Your Brain.*

24. Lisa Lewis, "Two Neuropsychological Models and Their Psychotherapeutic Implications," *Bulletin of the Menninger Clinic* 56 (1992), p. 28.

25. Kolb and Whishaw, p. 187.

26. William J. Hudspeth and Karl H. Pribram, "Stages of Brain and Cognitive Maturation," *Journal of Educational Psychology* 82, No. 4 (1990), p. 881.

27. G. Pawlik and W.-D. Heiss, "Positron Emission Tomography and Neuropsychological Function," in *Neuropsychological Function and Brain Imaging*, eds. Erin D. Bigler, Ronald A. Yeo, and Eric Turkheimer (New York: Plenum, 1989), p. 78.

28. Halgren, p. 201.

29. Kolb and Whishaw, p. 259.

30. Simon LeVay, *The Sexual Brain* (Cambridge, Mass.: MIT Press, 1993), p. 31.

31. John Bransford, Robert Sherwood, Nancy Vye, and John Rieser, "Teaching Thinking and Problem Solving," *American Psychologist* 41, No. 10 (October, 1986), pp. 1078–1089.

32. LeVay, p. 31.

33. Pawlik and Heiss, p. 75.

CHAPTER 4: THE HUNGRY AND BIASED BRAIN

1. Bryan Kolb and Ian Q. Whishaw, *Fundamental of Human Neuropsychology*, 4th ed. (New York: W. H. Freeman, 1996), p. 39.

2. Neil R. Carlson, *Physiology of Behavior*, 3rd ed. (Boston: Allyn and Bacon, 1986), p. 22.

3. M. C. Diamond, "Plasticity of the Brain: Enrichment versus Impoverishment," in *Television and the Preparation of the Mind for Learning*, eds. C. Clark and K. King, cited in Christine B. McCormick and Michael Pressley, *Educational Psychology: Learning, Instruction, Assessment* (New York: Addison Wesley Longman, 1997), p. 132.

4. R. L. Sidman and P. Rakic, "Neuronal Migration with Special Reference to Developing Human Brain: A Review," cited by Peter R. Huttenlocher, "Synaptogenesis in Human Cerebral Cortex," in *Human Behavior and the Developing Brain*, ed. Geraldine Dawson and Kurt W. Fischer (New York: The Guilford Press, 1994), p. 137.

5. Robert Y. Moore, "Normal Development of the Nervous System," in *Prenatal and Perinatal Factors Associated with Brain Disorders*, ed. John M. Freeman (U.S. Department of Health and Human Services, NIH Publication No. 895-1149, April 1985), pp. 47–48.

6. U.S. Department of Health and Human Services, *Decade of the Brain: Answers Through Scientific Research* (NIH Publication No. 88-2957, January 1989), Figure 1.

7. Kolb and Whishaw, pp. 39–40.

8. R. F. Schmidt, "The Structure of the Nervous System," in *Fundamentals of Neurophysiology*, ed. R. F. Schmidt (New York: Springer-Verlag, 1985), p. 1.

9. Carlson, p. 17.

10. Ibid., p. 19.

11. Kolb and Whishaw, p. 69.

12. Carlson, p. 23.

13. Henry Khachaturian, "Neurotransmission," in *The Neuroscience of Mental Health II: A Report on Neuroscience Research: Status and Potential for Mental Health and Mental Illness*, ed. Stephen H. Koslow (Rockville, Maryland: U.S. Department of National Institutes of Health, 1995), pp. 39–57.

14. Ibid.

15. James Bevan, *The Simon and Schuster Handbook of Anatomy and Physiology* (New York: Simon and Schuster, 1978), p. 58.

16. Khachaturian, p. 41.

17. Kolb and Whishaw, p. 79.

18. Ibid., p. 201.

19. Daniel T. Gilbert, "How Mental Systems Believe," *American Psychologist* 46, No. 2 (1991), p. 109.

20. Duane Schultz, *A History of Modern Psychology*, 3rd ed. (New York: Academic Press, 1981), p. 329.

21. Robert Ornstein, *The Evolution of Consciousness* (New York: Simon and Schuster, 1991), p. 132.

22. "You Can Put Brain on Autopilot," *Albuquerque Journal* (January 30, 1994), p. H3.

23. "Thought and Communication," in *Basic Behavioral Science Research for Mental Health: A National Investment* (Washington, D.C.: National Institute of Mental Health, NIH Publication No. 95-3682, 1995), p. 53.

24. Gilbert, p. 116.

25. Tina Adler, "By Age 4, Most Kids See What's Real," *APA Monitor* 22 (October 1991), p. 16.

26. Felicia Pratto and Oliver P. John, "Automatic Vigilance: The Attention-Grabbing Power of Negative Social Information," *Journal of Personality and Social Psychology* 61, No. 3 (1991), pp. 380–391.

27. "Social Influence and Social Cognition," *Basic Behavioral Science Research for Mental Health: A National Investment: A Report of the National Advisory Mental Health Council* (Washington, D.C.: National Institute of Mental Health, NIH Publication No. 95-3682, 1995), p. 73.

28. C. Neil Macrae, Alan B. Milne, and Galen V. Bodenhausen, "Stereotypes as Energy-Saving Devices: A Peek Inside the Cognitive Toolbox," *Journal of Personality and Social Psychology*, No. 1 (1994), p. 37.

29. "Social Influence and Social Cognition," p. 75.

CHAPTER 5: THE INDELIBLE STAMP OF DEVELOPMENT

1. Bryan Kolb and Ian Q. Whishaw, *Fundamentals of Human Neuropsychology*, 4th ed. (New York: W. H. Freeman, 1996), p. 42.

2. Robert T. Francoeur, *Becoming a Sexual Person* (New York: Macmillan Publishing Company, 1985), p. 206.

3. Christine B. McCormick and Michael Pressley, *Educational Psychology: Learning, Instruction, Assessment* (New York: Addison Wesley Longman, 1997), pp. 417–419.

4. Ibid.

5. Ibid., p. 119.

6. Peter R. Huttenlocher, "Synaptogenesis in Human Cerebral Cortex," in *Human Behavior and the Developing Brain*, eds. Geraldine Dawson and Kurt W. Fischer (New York: Guilford Press, 1994), p. 137.

7. Patricia S. Goldman-Rakic, "Development of Cortical Circuitry and Cognitive Function," *Child Development* 58 (1987), p. 615.

8. Ibid.

9. Harry T. Chugani, "Development of Regional Brain Glucose Metabolism

in Relation to Behavior and Plasticity," in *Human Behavior and the Developing Brain*, ed. Geraldine Dawson (New York: Guilford Press, 1994), p. 166.

10. Marc H. Bornstein, "Sensitive Periods in Development: Structural Characteristics and Causal Interpretations," *Psychological Bulletin* 105, No. 2 (1989), pp. 179–197.

11. Bruce D. Perry, "Incubated in Terror: Neurodevelopmental Factors in the 'Cycle of Violence,' " in *Children in a Violent Society*, ed. John D. Osofsky (New York: Guilford Press, 1997), pp. 124–149.

12. Sandra Blakeslee, "Re-evaluating Significance of Baby's Bond with Mother," *New York Times* (August 4, 1998), pp. 9+.

13. McCormick and Pressley, p. 133.

14. Ibid.

15. William J. Hudspeth and Karl H. Pribram, "Stages of Brain and Cognitive Maturation," *Journal of Educational Psychology* 82, No. 4 (1990), p. 881.

16. B. E. Wexler, "Cerebral Laterality and Psychiatry: A Review of the Literature," pp. 279–291, cited in Lisa Lewis, "Two Neuropsychological Models and Their Psychotherapeutic Implications," *Bulletin of the Menninger Clinic* 56, No. 56 (1992), p. 23.

17. *Decade of the Brain: Answers Through Scientific Research* (Washington, D.C.: National Institute of Health, NIH Publication 88-2957, January 1989), p. 22.

18. Perry, p. 8.

19. Ibid.

20. J. Piaget, *The Construction of Reality in the Child* (New York: Basic Books, 1954) translated by M. Cook, originally published in 1950.

21. Tina Adler, "By Age 4, Most Kids See What's Real," *APA Monitor* 22 (October 1991), p. 16.

22. Deirdre A. Kramer, "Post-Formal Operations? A Need for Further Conceptualization," *Human Development* 26 (1983), pp. 91–105.

23. John Bransford, Robert Sherwood, Nancy Vye, and John Rieser, "Teaching Thinking and Problem Solving," *American Psychologist* 41, No. 10 (October, 1986), pp. 1078–1089.

24. Bruce Perry, "The Effects of Early Childhood Trauma on Children, Families, and Communities," (Workshop Sponsored by the Family Centered Training Project, College of Education, University of New Mexico, Thursday, June 6, 1996, Albuquerque Convention Center).

25. "New Zealand Study Lasted 21 Years," *Prodigy Services Company* (July 21, 1994).

26. *Diagnostic and Statistical Manual of Mental Disorders*, 4th ed. (Washington, D.C.: American Psychiatric Association, 1994), p. 629.

27. Richard J. Kavoussi and Emil F. Coccaro, "Neurobiological Approaches to Disorders of Personality," in *Neuropsychiatry of Personality Disorders*, ed. John J. Ratey (Cambridge, Mass.: Blackwell Science, Inc.), p. 17.

28. Neil Swan, "NIDA Brain Imaging Research Links Cue-Induced Craving to Structures Involved in Memory," *NIDA Notes* 11 (November–December, 1996), p. 10.

29. Hilary Saner and Phyllis Ellickson, "Concurrent Risk Factors for Adolescent Violence," *Journal of Adolescent Health* 19 (August 1996), pp. 94–103.

CHAPTER 6: MEMORIES: LOYAL AND MISUNDERSTOOD

1. *American Heritage Dictionary: Based on the New Second College Edition* (New York: Dell Publishing, 1983), p. 428.

2. Henry Khachaturian, "Neurotransmission," in *The Neuroscience of Mental Health II: A Report on Neuroscience Research: Status and Potential for Mental Health and Mental Illness,* ed. Stephen H. Koslow (Rockville, Md.: U.S. Department of National Institutes of Health, 1995), pp. 39–57.

3. S. Zola-Morgan and L. R. Squire, "Neuroanatomy of Memory," *Annual Review of Neuroscience* 16 (1993), p. 547.

4. Neil R. Carlson, *Physiology of Behavior,* 3rd ed. (Boston: Allyn and Bacon, 1986), p. 591.

5. John Jonides, Edward E. Smith, Robert A. Koepper, Edward Awh, Satoshi Minoshima, and Mark A. Mintun, "Spatial Working Memory in Humans as Revealed by PET," *Nature* 363 (June 17, 1993), p. 623.

6. Marcus E. Raichle, "The Scratchpad of the Mind," *Nature* 363 (June 17, 1993), p. 583.

7. Jonides et al., p. 623.

8. H. Matthies, "Neurobiological Aspects of Learning and Memory," *Annual Review of Psychology* 40 (1989), p. 385.

9. "Essential Ingredient for Learning," (Prodigy Services Company, Reuters Wire Service, July 29, 1994).

10. David M. Diamond, Monika Fleshner, Nan Ingersoll, and Gregory M. Rose, "Psychological Stress Impairs Spatial Working Memory: Relevance to Electrophysiological Studies of Hippocampal Function," *Behavioral Neuroscience* 110, No. 4 (1996), p. 661.

11. L. R. Squire, B. Knowlton, and G. Musen, "The Structure and Organization of Memory," *Annual Review of Psychology* 44 (1993), pp. 453–495.

12. Khachaturian, p. 51.

13. Stephen B. Klein, *Learning: Principle and Applications,* 3rd ed. (New York: McGraw Hill, 1996), p. 503.

14. Bryan Kolb and Ian Q. Whishaw, *Fundamentals of Human Neuropsychology,* 4th ed. (New York: W. H. Freeman, 1996), p. 364.

15. Daniel L. Schacter, C.-Y. Peter Chiu, and Kevin N. Ochsner, "Implicit Memory: A Selective Review," *Annual Review of Neuroscience* 16 (1993), pp. 159–182.

16. Steven J. Sherman, "Social Cognition," *Annual Review of Psychology* 40 (1989), p. 282.

17. Joseph E. LeDoux, "Cognitive-Emotional Interactions in the Brain," *Cognition and Emotion* 3, No. 4 (1989), pp. 267–289.

18. Carlson, p. 630.

19. Deborah Chambers and Daniel Reisberg, "What an Image Depicts Depends on What an Image Means," *Cognitive Psychology* 24 (1992), p. 145.

20. Ibid.

21. Jean M. Mandler, "How to Build a Baby: II. Conceptual Primitives," *Psychological Review* 99, No. 4 (1992), p. 589.

22. Bruce D. Perry, "Incubated in Terror: Neurodevelopmental Factors in the

'Cycle of Violence,' " in *Children in a Violent Society*, ed. John D. Osofsky (New York: Guilford Press, 1997), pp. 127–128.

23. "Perception, Attention, Learning, and Memory," *Basic Behavioral Science Research for Mental Health: A National Investment: A Report of the National Advisory Mental Health Council* (U.S. Dept. of Health and Human Services, National Institute of Mental Health, 1995, Doc. HE 20.8102:B39/5 0507-B-05), p. 47.

24. Ibid.

25. Salvatore R. Maddi, *Personality Theories: A Comprehensive Analysis*, 3rd ed. (Homewood, Ill.: Dorsey Press, 1976), p. 191.

26. Duane Schultz, *A History of Modern Psychology*, 3rd ed. (New York: Academic Press, 1981), p. 329.

27. U.S. Department of Health and Human Services, *Decade of the Brain: Answers Through Scientific Research* (NIH Publication No. 88-2957, January 1989), p. 6.

CHAPTER 7: THE WELL CONDITIONED BODY

1. Michael Domjan and Barbara Burkhard, *The Principles of Learning and Behavior*, 2nd ed. (Monterey, Calif.: Brooks/Cole, 1986), p. 11.

2. Bryan Kolb and Ian Q. Whishaw, *Fundamentals of Human Neuropsychology*, 4th ed. (New York: W. H. Freeman, 1996), p. 363.

3. Robert W. Lundin, *Theories and Systems of Psychology*, 3rd ed. (Lexington, Mass.: D.C. Heath and Co., 1985), p. 169.

4. Patricia S. Goldman-Rakic, "Development of Cortical Circuitry and Cognitive Function," *Child Development* 58 (1987), p. 601.

5. Christine B. McCormick and Michael Pressley, *Educational Psychology: Learning, Instruction, Assessment* (New York: Addison Wesley Longman, 1997), p. 162.

6. David G. Lavond, Jeansok J. Kim, and Richard F. Thompson, "Mammalian Brain Substrates of Aversive Classical Conditioning," *Annual Review of Psychology* 44 (1993), pp. 317–342.

7. Duane Schultz, *A History of Modern Psychology*, 3rd ed. (New York: Academic Press, 1981), p. 192.

8. Ibid.

9. L. R. Squire, B. Knowlton, and G. Musen, "The Structure and Organization of Memory," *Annual Review of Psychology* 44 (1993), p. 477.

10. Robert A. Rescorla, "Pavlovian Conditioning: It's Not What You Think It Is," *American Psychologist* 43 (March 1988), p. 154.

11. "Perception, Attention, Learning, and Memory," in *Basic Behavioral Science Research for Mental Health: A National Investment* (Washington, D.C.: National Institute of Mental Health, NIH Publication No. 95-3682, 1995), p. 43.

12. Lundin, p. 168.

13. Ibid.

14. Ibid.

15. Rescorla, p. 153.

16. S. Zola-Morgan and L. R. Squire, "Neuroanatomy of Memory," *Annual Review of Neuroscience* 16 (1993), pp. 547–563.

17. David Gaffan, "Amygdala and the Memory of Reward," in *The Amygdala: Neurobiological Aspects of Emotion, Memory, and Mental Dysfunction*, ed. John P. Aggleton (New York: John Wiley, 1992), p. 471.

18. Kolb and Whishaw, p. 375; Robert J. McDonald and Norman M. White, "The Triple Dissociation of Memory Systems: Hippocampus, Amygdala, and Dorsal Striatum," *Behavioral Neuroscience* 107, No. 1 (1993), pp. 3–5.

19. J. Wolpe, *Psychotherapy by Reciprocal Inhibition* (Stanford: Stanford University Press, 1958), cited in *Handbook of Clinical Psychology: Theory, Research, and Practice*, Vol. 2, ed. C. Eugene Walker (Homewood, Ill.: Dow Jones-Irwin, 1983), pp. 968–969.

20. C. Eugene Walker, *Handbook of Clinical Psychology: Theory, Research, and Practice*, Vol. 2 (Homewood, Ill.: Dow Jones-Irwin, 1983), p. 537.

21. Ibid.

22. Stephen B. Klein, *Learning Principles and Applications*, 3rd ed. (New York: McGraw-Hill, 1996), p. 128.

23. *Managing Behavioral Challenges: Key to Student Achievement* (Denver Public Schools, Department of Student Services, 1998), Section 1, p. 3.

24. Klein, p. 204.

25. Ibid., pp. 205–212.

26. David R. Shanks, "Human Instrumental Learning: A Critical Review of Data and Theory," *British Journal of Psychology* 84 (1993), p. 322.

27. Domjan and Burkhard, p. 217.

28. Klein, p. 183.

29. Khachaturian, "Neurotransmission," p. 51.

30. Jeffrey M. Schwartz, Paula W. Stoessel, Lewis Baxter Jr., Karron M. Martin, and Michael E. Phelps, "Systematic Changes in Cerebral Glucose Metabolic Rate After Successful Behavior Modification Treatment of Obsessive-Compulsive Disorder," *Archives of General Psychiatry* 53 (Feb. 1996), p. 109.

31. Joseph E. LeDoux, "Emotion and the Amygdala," in *The Amygdala: Neurobiological Aspects of Emotion, Memory, and Mental Dysfunction*, ed. John P. Aggleton (New York: Wiley and Sons, 1992), p. 339.

32. Joseph E. LeDoux, "Cognitive-Emotional Interactions in the Brain," *Cognition and Emotion* 3, No. 4 (1989), pp. 267–289.

CHAPTER 8: SEEING OR HEARING IS DOING

1. Albert Bandura, *Social Learning Theory* (Englewood Cliffs, N.J.: Prentice-Hall, 1977).

2. Christine B. McCormick and Michael Pressley, *Educational Psychology: Learning, Instruction, Assessment* (New York: Addison Wesley Longman, 1997), p. 171.

3. "Social Influence and Social Cognition," in *Basic Behavioral Science Research for Mental Health: A National Investment* (Washington, D.C.: National Institute of Mental Health, NIH Publication No. 95-3682, 1995), p. 69.

4. Lisa Lewis, "Two Neuropsychological Models and Their Psychotherapeutic Implications," *Bulletin of the Menninger Clinic* 56 (1992), p. 26.

5. Ibid.

6. McCormick and Pressley, p. 173.

7. Ronald P. Philipchalk, *Invitation to Social Psychology* (New York: Harcourt Brace College Publishers, 1995), p. 206.

8. Ibid.

9. Muzafer Sherif and Carolyn Sherif, cited in Ronald P. Philipchalk, *Invitation to Social Psychology* (New York: Harcourt Brace College Publishers, 1995), p. 348.

10. Ibid.

11. Philipchalk, p. 355.

12. Ibid.

13. Sherif and Sherif, p. 356.

14. Brian M. Fagan, *People of the Earth*, 9th ed. (New York: Addison Wesley Longman, 1998), p. 17.

15. Robert J. Wenke, *Patterns in Prehistory: Humankind's First Three Million Years*, 2nd ed. (New York: Oxford University Press, 1984), p. 201.

16. David Sloan Wilson, "Human Groups as Units of Selection," *Science* 276 (June 20, 1997), p. 1816.

17. Willard W. Hartup, "Social Relationships and Their Developmental Significance," *American Psychologist* 44 (February 1989), p. 120.

18. Edward Z. Tronick, "Emotions and Emotional Communication in Infants," *American Psychologist* 44 (Feb. 1989), p. 112.

19. Bruce D. Perry, "Incubated in Terror: Neurodevelopmental Factors in the 'Cycle of Violence,'" in *Children in a Violent Society*, ed. John D. Osofsky (New York: Guilford Press, 1997), p. 132.

20. Tronick, p. 112.

21. Ibid.

22. Robert W. Levenson and Anna M. Ruef, "Empathy: A Physiological Substrate," *Journal of Personality and Social Psychology* 63, No. 2 (1992), p. 234.

23. Tronick, p. 112.

24. Ibid.

25. Ibid.

26. Claire B. Kopp, "Regulation of Distress and Negative Emotions: A Developmental View," *Developmental Psychology*, 25, No. 3 (1989), p. 343.

27. Tronick, p. 117.

28. Perry, p. 132.

29. Margaret S. Clark and Harry T. Reis, "Interpersonal Processes in Close Relationships," *Annual Review of Psychology* 39 (1988), p. 642.

30. Ross Buck, "Emotional Communication, Emotional Competence, and Physical Illness: A Developmental-Interactionist View," in *Emotion Inhibition and Health*, eds. Harold C. Traue and James W. Pennebaker (Seattle: Hogrefe & Huber Publishers, 1993), p. 226.

31. William J. Hudspeth and Karl H. Pribram, "Stages of Brain and Cognitive Maturation," *Journal of Educational Psychology* 82, No. 4 (1990), p. 881; Patricia S. Goldman-Rakic, "Development of Cortical Circuitry and Cognitive Function," *Child Development* 58 (1987), p. 602.

32. Newton N. Minow and Craig L. LaMay, *Abandoned in the Wasteland: Children, Television, and the First Amendment* (New York: Hill & Wang, 1995), pp. 3–15.

33. J. Decety, D. Perani, M. Jeannerod, V. Bettinardi, B. Tadary, R. Woods, J. C. Mazziotta, and F. Fazio, "Mapping Motor Representations with Positron Emission Tomography," *Nature* 371 (October 13, 1994), p. 600.

34. Wanda Wyrwicka, "Imitative Behavior," *Pavlovian Journal of Biological Science* 23 (July–Sept. 1998), p. 125.

35. Minow and LaMay, p. 5.

36. "Thought and Communication" in *Basic Behavioral Science Research for Mental Health: A National Investment* (Washington, D.C.: National Institute of Mental Health, NIH Publication No. 95-3682, 1995), p. 61.

37. Ibid., p. 62.

38. E. Newport, "Task Specificity in Language Learning? Evidence from Speech Perception and American Sign Language," cited in Michael Maratsos and Laura Matheny, "Language Specificity and Elasticity: Brain and Clinical Syndrome Studies," *Annual Review of Psychology* 45 (1994), p. 490.

39. C. Thomas Gualtieri, "The Contribution of the Frontal Lobes to a Theory of Psychopathology," in *Neuropsychiatry of Personality Disorders*, ed. John J. Ratey (Cambridge, Mass.: Blackwell Science, 1995), p. 150.

40. Goldman-Rakic, p. 604.

41. Bryan Kolb and Ian Q. Whishaw, *Fundamentals of Human Neuropsychology*, 4th ed. (New York: W. H. Freeman, 1996), p. 324.

42. Gualtieri, p. 150.

43. Richard E. Vatz and Lee S. Weinberg, "The Unabomber's Twisted Saga," *USA Today (Magazine)* 127 (July 1998), p. 53.

CHAPTER 9: THE RELIABLE BRAIN

1. Nico H. Frijda, *The Emotions* (Cambridge, Mass.: Cambridge University Press, 1986), p. 9.

2. Paul Ekman and Richard J. Davidson, eds., *The Nature of Emotion: Fundamental Questions* (New York: Oxford University Press, 1994), p. 5.

3. James R. Averill, *Anger and Aggression: An Essay on Emotion* (New York: Springer-Verlag, 1982), p. 4.

4. Daniel Goleman, *Emotional Intelligence* (New York: Bantam Books, 1995), p. 5.

5. Frijda, p. 12.

6. Ibid., p. 69.

7. Carol Tavris, *Anger: The Misunderstood Emotion* (New York: Simon & Schuster, 1982), pp. 15–288.

8. Ibid.

9. *Diagnostic and Statistical Manual of Mental Disorders*, 4th ed. (Washington, D.C.: American Psychiatric Association, 1994).

10. National Advisory Mental Health Council, "Basic Behavioral Science Research for Mental Health," *American Psychologist* 50 (October 1995), p. 838.

11. Klaus Schreiber, "The Adolescent Crack Dealer: A Failure in the Development of Empathy," *Journal of the American Academy of Psychoanalysis* 20, No. 2 (1992), p. 243.

12. Ibid.

13. Bessel A. van der Kolk, "The Body Keeps the Score: Memory and the Evolving Psychobiology of Posttraumatic Stress," *Harvard Review of Psychiatry* (Jan.–Feb. 1994), p. 253.

14. Frijda, p. 142.

15. J. D. Watt and M. J. Blanchard, "Boredom Proneness and the Need for Cognition," *Journal of Research in Personality* 28 (1994), p. 44.

16. Van der Kolk, p. 253.

17. Paul Ekman, "Moods, Emotions, and Traits," in *The Nature of Emotion: Fundamental Questions*, eds. Paul Ekman and Richard J. Davidson (New York: Oxford University Press, 1994), p. 56.

18. Ibid.

19. Ibid.

20. Frijda, p. 70.

21. Ibid., p. 380.

22. "Emotion and Motivation," in *Basic Behavioral Science Research for Mental Health: A National Investment* (Washington, D.C.: National Institute of Mental Health, NIH Publication No. 95-3682, 1995), p. 16.

23. Van der Kolk, "The Body Keeps the Score."

24. Joseph E. LeDoux, "Emotion and the Amygdala," in *The Amygdala: Neurobiological Aspects of Emotion, Memory, and Mental Dysfunction*, ed. John P. Aggleton (New York: John Wiley, 1992), p. 338.

25. Ibid.

26. Eric Halgren, "Emotional Neurophysiology of the Amygdala Within the Context of Human Cognition," in *The Amygdala: Neurobiological Aspects of Emotion, Memory, and Mental Dysfunction*, ed. John P. Aggleton (New York: John Wiley, 1992), pp. 191–228.

27. Ibid., p. 206.

28. Ibid., p. 212.

29. Frijda, p. 379.

30. LeDoux, p. 347.

31. "Emotion and Motivation," p. 14.

CHAPTER 10: TIME TRAVEL

1. James L. Gould and Peter Marler, "Learning by Instinct," *Scientific American* 256 (Jan. 1987), p. 74.

2. T. Berry Brazelton, "Joint Regulation of Neonate-Parent Behavior," in *Social Interchange in Infancy: Affect, Cognition, and Communication*, ed. Edward Z. Tronick (Baltimore, Md.: University Park Press, 1982), pp. 7–22.

3. Gould and Marler, p. 74.

4. Keith J. Holyoak and Barbara A. Spellman, "Thinking," *Annual Review of Psychology* 44 (1993), p. 278.

5. Lokendra Shastri and Venkat Ajjanagadde, "From Simple Associations to Systematic Reasoning: A Connectionist Representation of Rules, Variables and Dynamic Bindings Using Temporal Synchrony," *Behavioral and Brain Sciences* 16 (1993), p. 418.

6. Ibid.

7. Ibid.

8. Duane Schultz, *A History of Modern Psychology*, 3rd ed. (New York: Academic Press, 1981), p. 329.

9. Ibid.

10. Robert Ornstein, *The Evolution of Consciousness* (New York: Simon and Schuster, 1991), p. 132.

11. Lisa Lewis, "Two Neuropsychological Models and Their Psychotherapeutic Implications," *Bulletin of the Menninger Clinic* 56 (1992), p. 25.

12. Dean Falk, "Brain Evolution in Homo: The "Radiator" Theory," *Behavioral and Brain Sciences* 13 (1990), pp. 344–368, cited in Bryan Kolb and Ian Q. Whishaw, *Fundamentals of Human Neuropsychology*, 4th ed. (New York: W. H. Freeman, 1996), p. 33.

13. Susan T. Fiske, "Thinking Is for Doing: Portraits of Social Cognition from Daguerreotype to Laserphoto," *Journal of Personality and Social Psychology* 63 (1992), p. 877.

14. "Thought and Communication," in *Basic Behavioral Science Research for Mental Health: A National Investment: A Report of the National Advisory Mental Health Council* (Washington, D.C.: U.S. Dept. of Health and Human Services, National Institute of Mental Health), p. 53.

15. P. W. Cheng and L. R. Novick, "Causes versus Enabling Condition," *Cognition* 40 (1991), pp. 83–120, cited in Keith J. Holyoak and Barbara A. Spellman, "Thinking," *Annual Review of Psychology* 44 (1993).

16. "Thought and Communication," in *Basic Behavioral Science Research for Mental Health: A National Investment* (Washington, D.C.: National Institute of Mental Health, NIH Publication No. 95-3682, 1995), p. 55.

17. David Dunning, Dale W. Griffin, James D. Milojkovic, and Lee Ross, "The Overconfidence Effect in Social Prediction," *Journal of Personality and Social Psychology* 58 (1990), p. 568.

18. Susan T. Fiske, "Social Cognition and Social Perception," *Annual Review of Psychology* 44 (1993), p. 162.

19. T. S. Pittman and J. F. Heller, "Social Motivation," *Annual Review of Psychology* 34 (1987), pp. 461–489, cited in Susan T. Fiske, "Social Cognition and Social Perception," *Annual Review of Psychology* 44 (1993), pp. 155–194.

20. Dunning et al., p. 568.

21. Jerome Bruner, "The Narrative Construction of Reality," *Critical Inquiry* 18 (Autumn 1991), pp. 1–21.

22. Daniel Kahneman and Carol A. Varey, "Propensities and Counterfactuals: The Loser That Almost Won," *Journal of Personality and Social Psychology* 59, No. 6 (1990), pp. 1101–1110.

23. Mark D. Alicke, "Culpable Causation," *Journal of Personality and Social Psychology* 63, No. 3 (1992), p. 368.

CHAPTER 11: THE HUMAN MIND, THE GROUP MIND

1. Brian M. Fagan, *People of the Earth*, 9th ed. (New York: Addison Wesley Longman, 1998), p. 154.
2. Bruce D. Perry, "Incubated in Terror: Neurodevelopmental Factors in the 'Cycle of Violence,'" in *Children in a Violent Society*, ed. John D. Osofsky (New York: Guilford Press, 1997), pp. 124–125.
3. Workshop by William H. Polonsky, "Social Connections, Health & Longevity," (Palo Alto, Calif.: Health Science Seminars, March 6, 1998).
4. Kathleen McCarthy, "Belief in Paranormal Isn't OK, It's Harmful," *APA Monitor* 22, No. 11 (1991), p. 33.
5. Susan T. Fiske, "Social Cognition and Social Perception," *Annual Review of Psychology* 44 (1993), p. 156.
6. Keith Harrary, "The Truth about Jonestown: 13 Years Later—Why We Should Still Be Afraid," *Psychology Today* 25 (March–April 1992), p. 64.
7. David Sloan Wilson, "Human Groups as Units of Selection," *Science* 276 (June 20, 1997), p. 1816.
8. Robert J. Wenke, *Patterns in Prehistory: Humankind's First Three Million Years*, 2nd ed. (New York: Oxford University Press, 1984), p. 68.
9. Ibid.
10. Stephanie Coontz, *The Way We Never Were* (New York: Basic Books, 1992), p. 10.
11. Perry, p. 124.
12. "Child Abuse Report Released," *Prodigy Services Company* (April 3, 1996).
13. Coontz, p. 2.
14. Newton N. Minow and Craig L. LaMay, *Abandoned in the Wasteland: Children, Television, and the First Amendment* (New York: Hill and Wang, 1995), p. 5.
15. Nathan Seppa, "TV Displays Violence Without the Mess," *APA Monitor* (April 1996), p. 8.
16. Roy F. Fox, "Manipulated Kids: Teens Tell How Ads Influence Them," *Educational Leadership* 53, No. 1 (September), p. 77.
17. Paul Simon, "When Will We Run Out of Water?", *Rocky Mountain News Parade Magazine* [Denver] (August 23, 1998).
18. Perry, p. 124.
19. APA Monitor. "Middle-Aged Women Are More at Risk for Internet Addiction" (October 1996), p. 10.

INDEX